Automotive & Electrical Systems

Troubleshooting and Repair Basics

by
Vaughn D. Martin

Automotive & Electrical Systems

Troubleshooting and Repair Basics

by
Vaughn D. Martin

PROMPT®
PUBLICATIONS

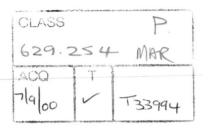
PROMPT©️ Publications is an imprint of Howard W. Sams & Company, A Bell Atlantic Company, 2647 Waterfront Parkway, E. Dr., Indianapolis, IN 46214-2041.

International Standard Book Number: 0-7906-1142-2
Library of Congress Catalog Card Number: 98-066672

Acquisitions Editor: Alice Tripp
Editor: Kim Heusel
Assistant Editor: Crystal Franklin
Typesetting: Kim Heusel
Cover Design: Christy Pierce
Graphics Conversion: Christy Pierce, Phil Velikan

PRINTED IN THE UNITED STATES OF AMERICA

9 8 7 6 5 4 3 2 1

Dedication

This is for the "Tin Wiz", Ron Keller, and Aldo Vasquez,
a man who rarely is satisfied with the degree
of precision of any mechanism.

Contents

Introduction

Automotive Electrical Systems—Troubleshooting and Repair Basics is for absolute beginners. It serves two purposes. First, if you have ever seen or tried to use a Chilton manual you know they are 1) very detailed and well illustrated, and 2) they cover a vast array of car models over a great number of years. However, they presuppose you already know enough to fully understand the subject. This book fills that *knowledge and experience* gap. Conquering this "gap" will allow you to naturally understand your car's electrical systems and computers well enough to buy, for example, replacement parts, and then a Chilton guide in which there are very detailed instructions on how to install this part.

Also, even if you are rather proficient at electronics, automotive schematics may still mystify you. Your car's electrical system looks far different than that of ordinary circuits represented in normal schematic form. For example, depending upon the manufacturer, a vacuum device can appear as a square box with a diagonal line(s) through it. This pictorial representation is by necessity, and represents an interaction between electrical and mechanical (pneumatic) components. The "necessity" is that, just as internationally recognized symbols such as the physically challenged wheelchair symbol are becoming very commonplace, automotive symbols are striving to do the same. This naturally boosts sales since it allows for a greater acceptance by non-English-speaking drivers.

Second, this book is for readers who feel that commercial garages charge too much for their services and parts, yet lack the skill to make their own car's electrical system repairs. How many times have you had your car repaired and the service writer quotes and charges you a price that exceeds the actual number of worked on your car? You probably ask yourself, "Was there more than one mechanic working on my car?" No. The service writer merely adds up individual task times, and charges accordingly; however, this is not an accurate estimate because the average experienced mechanic can perform most tasks far faster than the "average" times quoted from their standardized "labor hours" guidelines. This book makes you less hesitant to go to a wholesale car parts store. After reading this book, you'll feel confident enough to buy and change out your car's components. (Appendix B lists suppliers where you may purchase virtually every instrument or component this book covers).

Virtually every book written thus far within the automotive electronics field has the same outline. It starts with the basics of electricity, as we have done. However, it then delves into each of your car's major systems. These are typically your charging and discharging system, your cooling system, your ignition, fuel injection system, etc. This book stresses electronic test instruments over the more traditional approach of automotive systems. Although this book covers all these systems, it mainly stresses troubleshooting and techniques, plus how to use and interpret your test instruments' results. It does not subdivide the chapters on the basis of an automobile system(s) in the front part of this book. You are brought up to speed first on the basics and next on understanding how electronic test equipment works. This book does subdivide chapters on the basis of electronic test instruments. As an example, Chapter 2 introduces you to the DMM (Digital Multimeter), and its many features are covered while performing diagnostic and trouble-shooting tests. There is an emphasis on learning the equipment since the repair is greatly simplified once you have the knowledge and equipment with which to tackle it.

The first chapter deals with basic electricity. The second chapter starts to examine one of your car's main systems, the cooling system, and first introduces you to a DMM to diagnose possible cooling system faults. Admittedly, while your cooling system is one of the least "electronic" in nature systems in your car, this same system simplicity serves us well since it presents an ideal and painless starting point. You will make a few simple measurements (such as checking your radiator temperature and thermostat) on your cooling system, followed by more involved troubleshooting tasks with a DMM. The third chapter covers more involved and advanced applications and measurements you can make with your DMM. This generally helps to graphically illustrate the powerful nature of electronic test equipment. The fourth chapter covers DSOs in theory and practice. This chapter also covers a combination of a DSO and a multimeter. It is made by Fluke and Tektronix (among others) and is quite appropriately called a scopemeter, which it is! The fifth chapter covers your car's many sensors and how you even use commercially available sensors to perform such vital tasks as measuring vibration and rotational motion of a part within your car's engine or drive train. You do this almost exclusively with a DSO (digital storage oscilloscope).

Chapters 6 and 7 assume the form of the more "traditional" books on automobile electronic troubleshooting that are divided on the basis of your car's systems. Chapter 6 shows how you can use oscilloscopes to make electrical measurements on your car's electrical systems. Chapter 7 begins the coverage of your car's computer. Since it comprises so much, such as alternator and voltage regulator etc. it is given a disproportionately large amount of coverage.

Appendix A gives the answers to all the questions which appear at the end of each chapter.

This book has numerous test procedures. Some of them overlap or repeat one another; however, despite being identical tasks, you perform them with different test equipment. This requires you to know how to interpret the same results in different visual formats. But this is not a problem since there is ample coverage of what test results should look like and how you can correctly evaluate them. This counterintuitive approach is purposeful and intentional. This is because it instructs you in more than one way, since you may not have one piece of test equipment but just may have another.

Large companies specializing in this business manufacture a great variety of automotive electronics test equipment. As an example, Fluke Corporation, or just Fluke, was recently bought by the parent company of Snap-On Tools. Many of these companies can give you attention in the form of application notes and advice sometimes gained through an 800 or 888 number (when one exists). They usually do their best if you are trying to use their test instrument. Wherever possible, this book mentions actual brand names, manufacturers, and part numbers of test equipment, as well as car components. This tends to make the procedure simpler, far more "hands-on," and practical.

These four types of electronic test instruments have variations. As an example, under the digital multimeter types, there are conventional ones, specialized ones dedicated just for cars, and even ones with a graphical very literal (icon presentation) LCD display. This last type may prove to be the most "user friendly" to the novice. While following the various troubleshooting and repair procedures, you should realize that—regardless of the test instrument you use—it is the actual procedure, and adjustments, in which you should be most interested.

1

Understanding
Basic Electricity

So often you hear that new cars are **so sophisticated** *that they defy anybody, except a well-equipped garage mechanic, to work on them. The purposeful design and intent of this book is to show you just how wrong this is. Admittedly, your car's electrical and computer systems are unquestionably the "Achilles heel" or the most-likely-to-fail systems in your modern car. In fact, a* **Consumer's Report** *survey of 425,000 cars even reported that half of all repairs were performed on these very prone-to-fail systems. This survey has great credibility since it included feedback from both car owners and garages. In the future, cars will only have more sophistication. This will mostly involve added electronics, sensors, emission controls, and on-board computers.*

Dispelling an Utter Myth

This promise of "more bad things to come" should not alarm you. You can still fix modern cars if you have the proper knowledge and a few handy tools and instruments. That's precisely what this book is all about. Things are only *bad* if you don't understand them. But since you are reading this book right now, you definitely don't need any further convincing of this! At least you doubt some of this commonly held myth of "too complex to work on these days."

The Simple-to-Complex "Divide and Conquer" Approach

This book's game plan is simple. We are dealing with electricity in a practical fashion, which is essentially what this book is about. Therefore, the first chapter examines Ohm's Law. This is the very underpinning of electricity. By using Ohm's Law, you start with the most fundamental concepts. You will then work your way up from there. Ohm's Law is absolutely vital to understanding electricity. It relates electricity's three quantities to one another. These are voltage, current and resistance. As you progress, don't be too concerned if there is any part of this chapter you do not fully understand. It will become clearer to you through actual practical troubleshooting examples in later chapters. In real life, and also within this book, there are many generalized procedures. They are not difficult, nor that complex. You will be able to easily follow them. Later you will even use and adapt them to fix your specific car.

This chapter covers eight subjects.

1. The all-important Ohm's Law.
2. Series, parallel and series-parallel circuits.
3. Charge and the states of energy.
4. Batteries and EMF (Electromotive Force).
5. Capacitance.
6. Inductance.
7. AC or alternating current, something your alternator produces. It gets changed (rectified) into DC or direct current after that. This is the only form of current your car battery can use or produce itself.
8. Your safety while working on your car and while using electronic test instruments.

It's All Based on Ohm's Law—There Are No Exceptions to This Rule!

If you are an absolute beginner in electronics, the best way to start is by *visualizing* electricity. You do this by comparing it to something you can see, touch, and, therefore, far more easily understand. From there, you can draw analogies and your own conclusions. This approach helps you better understand this ***illusive*** invisible quantity called electricity. A simple equation (Eq. 1-1) represents Ohm's Law. Again, this is the basis for describing all electricity and electrical circuit flow. There are only three quantities involved. These are voltage in volts, current in Amperes, and resistance in ohms. You may also sometimes see resistance expressed in Ohms with a Greek letter (Omega). This is it: Ω.

The Water Tower Analogy

There is a classical analogy which fits electricity very well. It is a water tower filled with water and having a faucet controlling the rate of water flow (see Fig. 1-1.) The height of the water gives **push** that equates to voltage. The faucet, and how open it is, provides varying degrees of **resistance** to the flow of the water. This faucet setting determines the resistance to the water flow. The actual **flow** of water naturally equates to current flow. By using both everyday and electrical terms, you will be able to visualize this relationship. You will do so in terms of three familiar concepts:

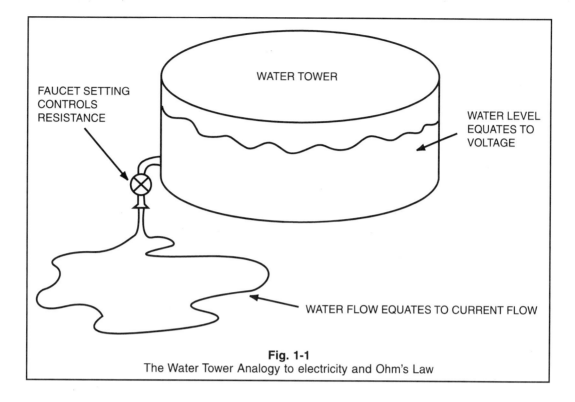

Fig. 1-1
The Water Tower Analogy to electricity and Ohm's Law

1. Push is voltage.
2. Resistance to water flow is resistance to electrical current flow.
3. The actual water flow equates to current flow.

Defining Electricity's Letter Symbols

The following, Equation 1-1, is Ohm's Law stated in the simplest way possible,

Voltage in Volts = Current in Amperes / Resistance in Ohms

Stated differently: **E = I x R** or **I = E / R** or **E = I x R**

E is voltage and it stands for **E**lectromotive force, thus the letter **E**. **I** stands for current and it got that **I** convention from the French of our English word intensity. Kind of confusing, isn't it?

There's an easier way

Memorizing anything, especially formulas, is the worst way to learn. So let's take an alternative approach. Figure 1-2 shows you a pie chart composed of **E, I,** and **R** or voltage, current, and resistance, respectively. If you want to find any one of these, e.g., current (**I**), just place your thumb over **I**. This leaves **E / R**. That means that current is voltage divided by resistance. Now let's check our water tower analogy. This means water's push divided by the

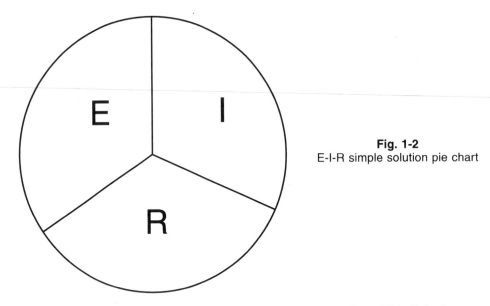

Fig. 1-2
E-I-R simple solution pie chart

faucet's resistance to water flow equals electrical current or water flow. This "placing your thumb over the unknown quantity" holds true for finding all three of Ohm's Law's electrical quantities.

First Things First—DC Current

A battery produces DC (Direct Current). This means that the current flows in one direction only. It does not at any time reverse its path in the opposite direction. Current flow begins at the positive 12 VDC (Volt Direct Current) and goes through all of the loads in the loop until it reaches your battery's negative terminal. (Fig. 1-3.) This negative terminal is called ground. It has a schematic or symbolic representation of a vertical line equally intersecting a horizontal line with two small horizontal lines below that (see Fig. 1-3 again). Almost all devices in your car only run off of DC (Direct Current).

An Example of Ohm's Law

Let's go through a typical simple example. We'll use electrical components and Ohm's Law to gain an appreciation and better understanding of the relationship between these three quantities. For clarity, although Amperes is the proper term for current, common use has shortened it to just Amps, or milliAmps which are 1/1,000 of an Amp. See the inside of this book's cover for these suffixes and prefixes. You use these to describe very small and large quantities. These are far larger or smaller than normal electrical symbols alone, such as Farads, Amps, Ohms, but rarely volts. You will note that all these prefixes and suffixes are to a power of 3, or some multiple thereof. This is called scientific notation. You will find this on many scientific calculators. When you do, note that all exponents will be expressed to a power of three, whether they are negative or positive!

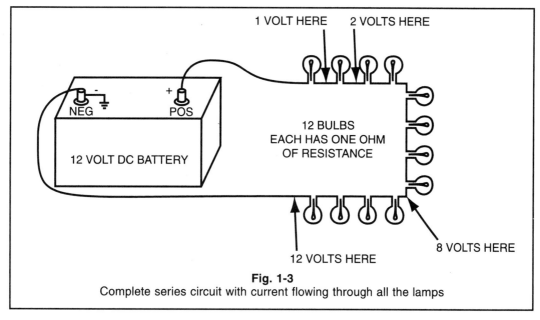

Fig. 1-3
Complete series circuit with current flowing through all the lamps

We will use Amps mostly from now on. In practice, a low resistance to either water or current flow means that there will be a larger amount of gushing water or current flowing. This condition occurs in situations with components that have small resistances. Conversely, a high-resistance component value causes less current to flow. Since current must flow through each device (a resistor representing a bulb in this case) before completing a circle (circuit), it has to go back to the ground terminal or where a zero volts potential exists. (Fig. 1-3.) The + terminal is naturally +12 VDC or Volts DC (Direct Current).

Note that the meter has a moving hand. This is the most prominent distinguishing characteristic of an analog meter. A digital meter has no moving hand. Figure 1-4 compares the readouts of these two types of meters. We will mostly use the more modern type of meter, the DMM (Digital Multimeter) in our work (Fig. 1-5). But more on these instrument matters later when we devote considerable text explaining how each works, its merits, and its disadvantages.

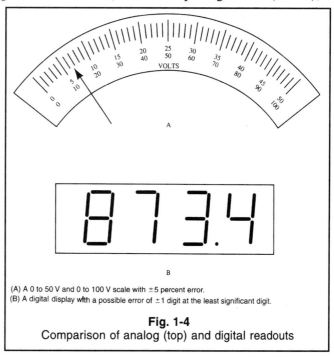

(A) A 0 to 50 V and 0 to 100 V scale with ±5 percent error.
(B) A digital display with a possible error of ±1 digit at the least significant digit.

Fig. 1-4
Comparison of analog (top) and digital readouts

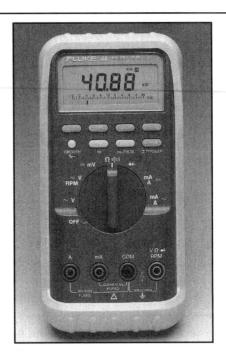

Fig. 1-5
Typical commercially available digital
multimeter. (Courtesy Fluke)

Figure 1-3 shows a car battery driving 12 identical bulbs. Each bulb has 1 Ohm of resistance. From this illustration, it should now be obvious that all components are in series. When you first solve for current, you arrive at 1 Amp. Next, if you are wondering about voltage drops along the circuit, you need to first relate voltage to the other two quantities in Ohm's Law. You will discover than $E = I \times R$. So we know I now and the R of each bulb is 1 Ohm. Therefore 1 Amp x 1 Ohm = 1 Volt. So there will be a 1-volt voltage drop across each of the 12 bulbs. The illustration in Fig. 1-3 clearly shows this.

Another Example

Assume you turn on your car's lights and they draw 2.0 Amps. Your car has a 12V battery. How much resistance does the wiring to these lights and the filaments within your car's lights present? We will soon see that wiring provides so little resistance in electrical circuits within your car that it is negligible. Therefore, we are just concerned with the filaments or lighting elements. These light sources can also be gases that vaporize, such as halogen, and illuminate. When enough current flows through the gas it breaks down and actually changes state. When that happens it gives off a bright light. Using Ohm's Law $E = I \times R$,

You are solving for **R**:

Since you know the voltage (12 volts) and current (2 Amps), you can easily and correctly find the missing or third item (Resistance) in Ohm's Law:

12 volts = 2 Amperes x X (expressed in Ohms or Ω) then X becomes 6 Ohms (Ω)

Further Defining an Electrical Circuit

Before we go any further, you need to understand what we mean by an electrical circuit. A circuit occurs when you interconnect electrical components in series. These components must allow current to flow through all of them. After this, current then flows back around to the battery's negative or ground terminal. You already know the schematic symbol for ground. The symbol in Fig. 1-3 on the battery's negative terminal represents a *ground* in schematic form in case you have forgotten. In your car, this voltage source is usually your 12-volt battery. As a practical matter, your car battery does not run down like the water height does within our water tower analogy. Your car has an alternator, which replenishes its energy (to be discussed later.) The sources of resistances or components within this *series* electrical circuit, when in use, draw current from the voltage source (your car's battery). Collectively, you call these components, which are energy consumers, **loads.**

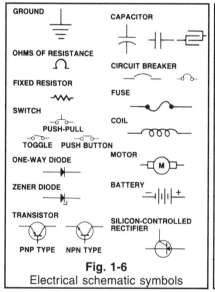

Fig. 1-6
Electrical schematic symbols

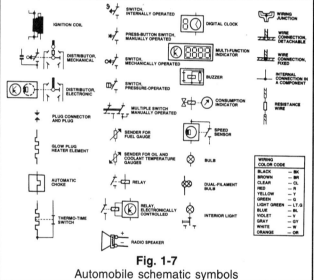

Fig. 1-7
Automobile schematic symbols

Schematic Symbols

The "loops" correspond to the filaments. The interconnecting lines in this drawing represent this particular portion of your car's electrical wiring system. When you see two lines cross, sometimes there is a dot or period at this intersection. That indicates a connection exists there. This connection can be a soldered joint or two or more wires crimped together. You do this by stripping back the wire's plastic insulation to where it's bare. Then you crimp (clamp) or hold these wires firmly and permanently in place. In fact, Figs. 1-6 and 1-7 compare both electrical and automotive symbols.

A Practical Example of A Series Circuit

Current is common to all components in a series circuit. Figure 1-3 is a typical series circuit. The twisted loop-appearing objects within the small circles represent incandescent light bulbs. Now go back to Fig. 1-3 and let's again realize that these filaments are what get hot. In fact, they get so hot when current flows through them that they glow. The amount of current flowing through them is proportional to their emitted light. However, there is a limit, and this is always specified when you buy these types of parts. These filaments are very simple and practical examples of components with resistances. All of your car's components have some resistance. But some components have more than others, as we will see next.

The "Expanded" Concept of A Series Circuit

As we just saw, current flows through all of a series circuit's components, and back to the voltage source. The amount of current flowing through each component is the same. That is what we mean by current is *common* to a series electrical circuit's components (or loads). So if you have eight loads connected in series, the current through any single load is the same as that of any other load. You can select the first and last loads or any combinations and this holds universally true in a *series* circuit only. This is what constitutes a full loop or circuit. Another law here relating to series circuits is that whatever your voltage source is, e.g., 12 volts, the sum of all these voltage drops must equal that same voltage source. In this case, it is the 12V battery in your car. This accounting for all the voltages (voltage drops) is vitally important. You will constantly use this in troubleshooting your car's electrical system, especially when you use a voltmeter.

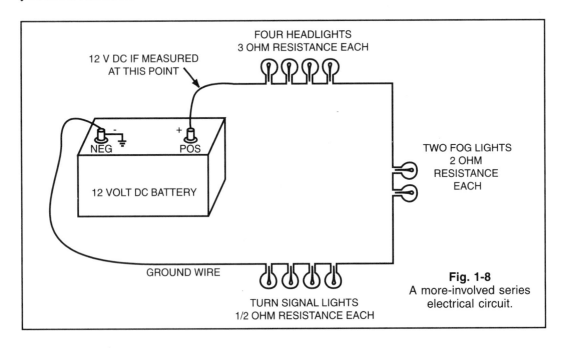

Fig. 1-8
A more-involved series
electrical circuit.

Let's take an example to demonstrate this (Fig. 1-8). Let's assume your car has four headlights, all identical, with 3 Ohms of resistance each. Then you have four smaller turn signal lights with one-half Ohm each of resistance. Lastly, you have two fog lamps with 2 Ohms of resistance each. If all these lights are in series and on at the same time, how much current flows through this circuit? Or stated differently, how much current does your battery have to supply? Also, how much of a voltage drop occurs across each of these eight lights? You first determine the voltage drops by finding the circuit's overall current. To check ourselves, once we have solved both for overall current and the total of all voltage drops, we'll show how these two answers relate to each other. It may already be apparent to you, but let's do the numbers nonetheless. Since I = E / R, and we already know E and R, we must solve for I. All you have to do is set up E and R. That is:

I = 12 volts / R and R = (4 x 3 Ohms) + (4 x ½ Ohm) + (2 x 2 Ohms) = 18 Ohms

12 volts / 18 Ohms = 2/3 or 0.667 Amps

Since 2/3 Amps flows through this string of 10 series connected lights, how much of a voltage drop occurs across just the four headlights altogether? Since you are solving for a voltage drop, you will use Ohm's Law to find this voltage.

This is: E = I x R or E = 0.667 Amps x (4 x 3 Ohms) = 0.667 Amps x 12 Ohms = 8 volts drop across the four headlights. The headlights take the lion's share of power, as they should, since they are far brighter than your car's other lights combined.

Let's trace the voltage drops across each component in Fig. 1-8. A crucial point that can not be overstressed is that all of a circuit's electrical components experience a voltage drop. The **sum of all these drops must equal the voltage source**. Your car's 12V battery is the source in this case. As you recall from Ohm's Law, current in Amps multiplied by resistances in Ohms = Voltage in volts. The first step is to determine current. This is the common parameter in a series circuit. Using Ohm's Law and referring to Fig. 1-8, you will note that we have a total of 18 Ohms of resistance and your car battery's 12-volt power source.

Series, Parallel, and Series-Parallel Circuits

Before moving to more complex circuits, let's redefine a series circuit one last time. We do so because it is so crucial to your understanding of all that follows. A series circuit must have three characteristics. It cannot be a series circuit unless it has all three of these characteristics. These characteristics are:

1. Current must flow through all components in the circuit—from the positive back around to the negative (ground) terminal.
2. Current in the circuit must be uniform and the same as it flows through each component— even if they have different resistances.
3. Voltage drops in a series circuit must be proportional to the resistance each component presents to current flow. The term proportional means in its respect to all of the components in the circuit.

Parallel Circuits

These use the same voltage source—the 12V battery in your car. But the components in a parallel circuit are arranged differently. They appear across your battery or voltage source. Figure 1-9 shows a typical parallel circuit. In this example, there are 12 volts across each lamp. But here is where it gets tricky. You recall how simple it was to determine current in a series circuit. Well, it is a little harder here. The current splits equally between these two lamps, each of which have

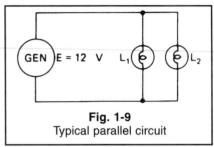

Fig. 1-9
Typical parallel circuit

an equal 60 Ohms of resistance. But before you can determine the amount of current which should flow, you'll have to simplify these two loads (resistors to current flow, actually lamps). The formula for this is: R1 + R2 / [R1 x R2] = R total. So if both loads are 60 Ohms, you transform their two resistances into one effective resistance (one single lamp). This would be

60 + 60 / [60 x 60] = 30 Ohms. In fact, when you have two identical loads in parallel, you can quickly assume that their effective resistance is just one-half of each one. That means that the total current flow in them is as if it were a single 30 Ohm lamp. By being able to find the resistance this way, you can then treat the parallel circuit like a series circuit. But have you noticed anything else? Throughout

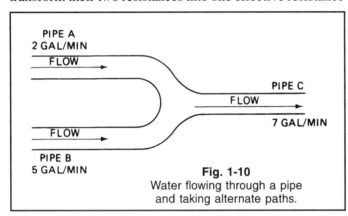

Fig. 1-10
Water flowing through a pipe
and taking alternate paths.

this analysis, voltage is the common element across these lamps. Contrast this to a series circuit in which current is common to all its components.

Returning to our water flow analogy, Fig. 1-10 verifies this in another way. This is that water flows through multiple sources and therefore splits (takes parallel paths). The resistance presented by each resistive element controls the amount of current flow. ***Current***

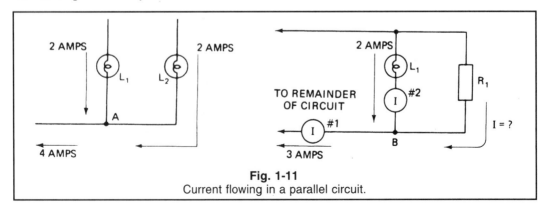

Fig. 1-11
Current flowing in a parallel circuit.

flow seeks the path of least resistance. Therefore, the total resistance of a parallel circuit is less than the smallest branch resistance. You would therefore expect a parallel branch with 2 Ohms of resistance to draw twice as much as a parallel branch with just 2 Ohms of resistance, since its total resistance is 1 Ohm. Figure 1-11 (left illustration) represents this current-splitting effect. There are equal resistances; therefore, there are equal currents flowing and the sum of these two currents equals the total current flowing within this parallel circuit. The illustration to the right in Fig. 1-11 shows a missing, or unidentified current. Since the sum of all currents must equal the total current, we have a total of 3 Amps, and one branch has 2 Amps. That means the remaining parallel branch with an $I = ?$ (the unknown value) has 1 Amp of current flowing through it.

The Concept of a Switch

If you were to solder each component permanently in place in a series circuit it would work. But when you turned your car off, these components would continue to drain your battery. What you need is a method of connecting and disconnecting components within a circuit. This is precisely what a switch does. It is a mechanical device consisting of two conductors (contacts or just plain wire segments) which the switch either unites with intimate contact or physically separates. The closed and open positions of the switch respectively represent these two possible situations. Quite simply, a closed switch allows current flow. An open switch blocks or prevents current flow.

Now let's look at several actual circuits to clarify this. Figure 1-12 shows a simple switch in the circuit in the upper left and it must close before current flows. The circuit to the right of this one must have two switches closed before current flows. These are both series circuits. The circuit to the lower left has two switches in par-

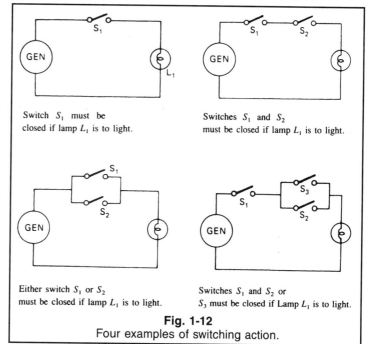

Switch S_1 must be closed if lamp L_1 is to light.

Switches S_1 and S_2 must be closed if lamp L_1 is to light.

Either switch S_1 or S_2 must be closed if lamp L_1 is to light.

Switches S_1 and S_2 or S_3 must be closed if Lamp L_1 is to light.

Fig. 1-12
Four examples of switching action.

allel. But if either is closed, the current flows. The remaining circuit to the lower right has a series switch, S1, and two switches in parallel, S2 and S3. You can get far more complex than this. Figure 1-13 shows two parallel switches. They are hooked in series. One of the dual switches in switch S1/S2 and switch S3/S4 must both be closed for current to flow. The other

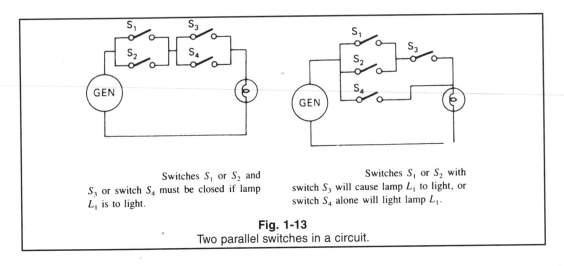

Switches S_1 or S_2 and S_3 or switch S_4 must be closed if lamp L_1 is to light.

Switches S_1 or S_2 with switch S_3 will cause lamp L_1 to light, or switch S_4 alone will light lamp L_1.

Fig. 1-13
Two parallel switches in a circuit.

circuit in Fig. 1-13 has a parallel switch in series with another parallel switch, plus one of the switches in switch S4 can bypass switch S3 if you close it.

Batteries

Your battery is the very "engine" or prime mover within you car's electrical system. Batteries have been around for more than 200 years. However, it has been only recently in this long history that their management (charging and discharging) is undergoing a mild revolution. Forces making them more environmentally friendly predominantly drove these changes. There are now ICs (integrated circuits) that help manage these two vital functions in your car's battery. Historically, Allesanaro Volta, probably unknowingly, discovered the battery in 1789. He placed copper and zinc rods into a vat of acetic acid. Naturally, the acid began to immediately eat away at these two metal rods. But more importantly, the copper rod captured the released energy in the form of ions. The copper and zinc rods served as positive and negative electrodes, respectively. Keeping that in mind, see Fig. 1-14. This shows a typical outline of a battery immersed in acid.

A Car Battery's Internal Resistance

This is an important factor in applications in which you draw large amounts of current over a short interval. This precisely describes the operating scenario of your car's electrical system—especially when you initially start your car. It becomes even more important if the outside temperature drops often to below 0°F (see Figure 1-15). The best analogy for this is to imagine a series resistor in line with your battery. This resistance multiplies by the current flowing through this series connection. The sum, in volts, is the amount of volts dropped (wasted). This wasted voltage does not contribute to the overall starting power your car desperately needs, especially at colder temperatures.

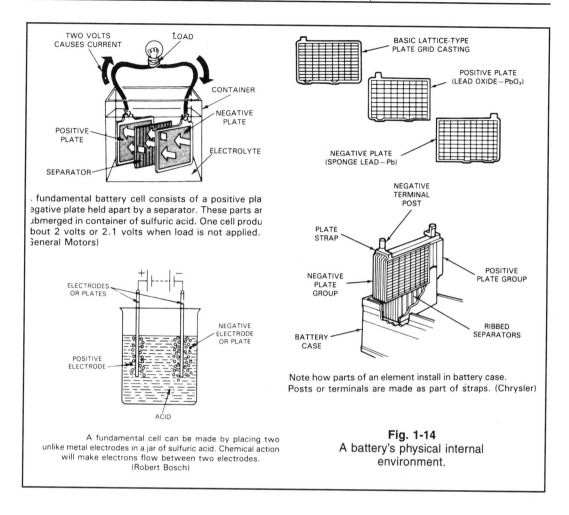

TWO VOLTS CAUSES CURRENT · LOAD · CONTAINER · NEGATIVE PLATE · POSITIVE PLATE · ELECTROLYTE · SEPARATOR

. fundamental battery cell consists of a positive pla
əgative plate held apart by a separator. These parts ar
ıbmerged in container of sulfuric acid. One cell produ
bout 2 volts or 2.1 volts when load is not applied.
ɔeneral Motors)

BASIC LATTICE-TYPE PLATE GRID CASTING · POSITIVE PLATE (LEAD OXIDE—PbO₂) · NEGATIVE PLATE (SPONGE LEAD—Pb)

NEGATIVE TERMINAL POST · PLATE STRAP · NEGATIVE PLATE GROUP · POSITIVE PLATE GROUP · BATTERY CASE · RIBBED SEPARATORS

Note how parts of an element install in battery case.
Posts or terminals are made as part of straps. (Chrysler)

ELECTRODES OR PLATES · NEGATIVE ELECTRODE OR PLATE · POSITIVE ELECTRODE · ACID

A fundamental cell can be made by placing two
unlike metal electrodes in a jar of sulfuric acid. Chemical action
will make electrons flow between two electrodes.
(Robert Bosch)

Fig. 1-14
A battery's physical internal
environment.

A frayed battery cable is a major contributing factor causing **I** x **R** or voltage drops. Figure 1-16 actually shows a host of causes which all contribute to poor starting problems. Assume that half of the batter cable has been cut. That means its resistance is doubled. This robs even more voltage and current from turning over your car's engine. You can measure your battery's internal resistance with a 0.01-Ohm resistor and a dampened ammeter (Figure 1-17). A dampened ammeter is a purposely-sluggish meter. Its indicator hand or pointer does not fly over to the left so hard that it literally wraps itself around the meter's left stopping post. Figure 1-18 shows an ammeter with a shunt. This is a resistor with low resistance that provides a parallel path for current. If this current division is 1,000 to 1, then the reading you observe is correct, but the meter actually only experiences 1/1,000 of the total current it displays.

Figure 1-19 shows two types of batteries. One has its terminal on top, and the other on its sides. You have no doubt seen these batteries many times when you opened your hood. There are four areas for immediate visual inspection you can perform. These are:

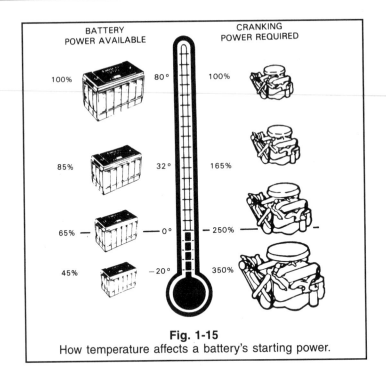

Fig. 1-15
How temperature affects a battery's starting power.

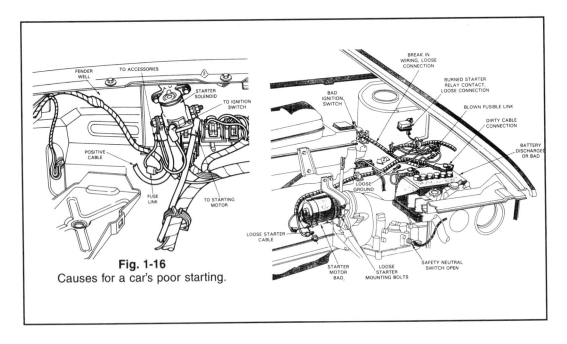

Fig. 1-16
Causes for a car's poor starting.

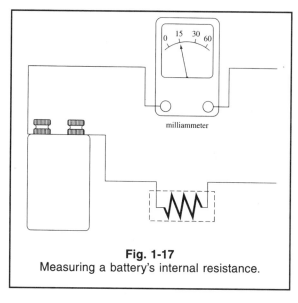

Fig. 1-17
Measuring a battery's internal resistance.

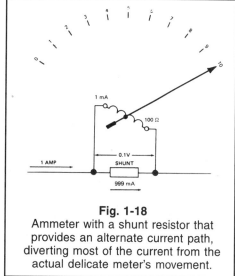

Fig. 1-18
Ammeter with a shunt resistor that
provides an alternate current path,
diverting most of the current from the
actual delicate meter's movement.

1. Dirt buildup (dirt is normally an insulator) around your battery case's top.
2. Damage to the battery's case (typically a cracked case).
3. Loose or corroded connections to and from your battery.
4. Rusted, or missing altogether, battery hold-down brackets or fixtures.

These adverse circumstances provide an undesirable unintended path for current flow, or ***battery leakage.*** Figure 1-20 shows how you test battery leakage. First, set your ohmmeter to its lowest scale. Then, touch points 1 and 3, the battery's terminals. Next, test from test point 1 to 2, 3 and 4 and note if your readings are different. That is, there should be no more than 0.25 volts across the test points, with the exception of the + and – terminals. Here, it should register an absolute minimum of 12.6 volts, or higher. Preferably it should be in the 13- to 13.5-volt range as you test across these + and – terminals. If the reading is low, you need to recharge your battery.

Other Battery Problems

If you have to frequently add water to your battery, you are overcharging it. Overcharging causes excessive current to enter your battery. The battery relieves itself by generating gas and venting it to the outside. (See the upper left area of Figure 1-16.) The inter-illustration caption "overcharged acid less" signifies this condition. This transformation of water-to-gas is also how your battery liberates itself of excessive water. The generated and transformed gas exits your battery through a vent. Virtually all batteries have a vent.

Battery Leakage Tests

As previously stated, dirt is normally an insulator. However, when there are oils, greases, and small metal fragments or filings mixed in, this dirt-based combination becomes a conductor. If there is a span of dirt from the battery's + to – terminals a connection exists through this

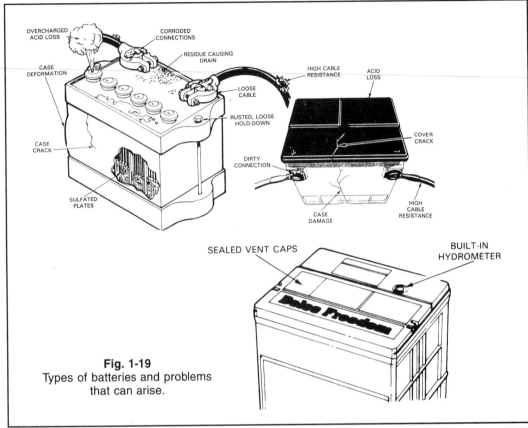

Fig. 1-19
Types of batteries and problems
that can arise.

dirt. This can exist by any number of means. Its mere presence establishes an unintended drain or conducting path for "lost and wasted" current.

Checking Battery Charge

When you check your battery's charge, you are actually checking the status of its *electrolyte*. An electrolyte is your battery's acid. It is a mixture of sulfuric acid and distilled water. It functions as a liquid medium by which ions are transported from the + to – terminals. Sulfuric acid is a very powerful corrosive substance. Avoid touching this or accidentally spilling it on your car, or rust will surely set in. Figure 1-21 shows three ways in which you

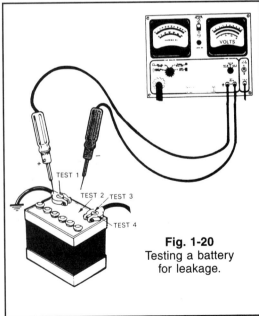

Fig. 1-20
Testing a battery
for leakage.

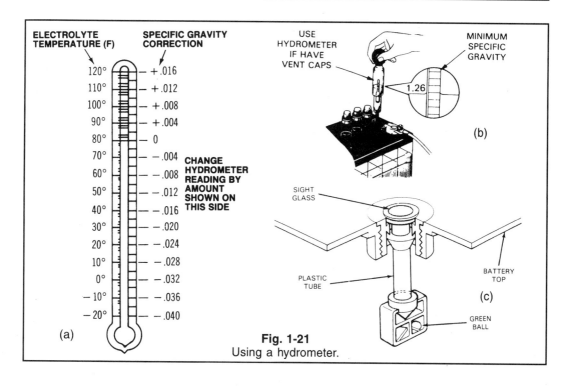

ELECTROLYTE
TEMPERATURE (F)

SPECIFIC GRAVITY
CORRECTION

120° — +.016
110° — +.012
100° — +.008
90° — +.004
80° — 0
70° — −.004 **CHANGE**
60° — −.008 **HYDROMETER**
READING BY
50° — −.012 **AMOUNT**
SHOWN ON
40° — −.016 **THIS SIDE**
30° — −.020
20° — −.024
10° — −.028
0° — −.032
−10° — −.036
−20° — −.040

(a)

USE
HYDROMETER
IF HAVE
VENT CAPS

MINIMUM
SPECIFIC
GRAVITY

1.26

(b)

SIGHT
GLASS

PLASTIC
TUBE

BATTERY
TOP

(c)

GREEN
BALL

Fig. 1-21
Using a hydrometer.

can check your battery's electrolyte. Figure 1-21(a) shows using a *hydrometer*. This instrument measures the specific gravity of a liquid. Specific gravity is the measure of the weight of the sulfuric acid to the weight of water. This ratio is only valid when measuring equal volumes of both liquids. Sulfuric acid is heavier than water. Therefore, it is not surprising to discover that the optimum specific gravity of your battery's electrolyte is from 1.265 to 1.299. The second method (Fig. 1-22) is easier. You merely observe the visual indictor tops of your battery's cells. In fact, most modern car batteries do not have removable tops from which you can gain access to the electrolyte with a hydrometer. These color codes have a purpose, as Fig. 1-22 illustrates.

Returning to our frayed battery cable problem, you can repair this. You strip back the insulation until you expose the bare wire. You solder the cable after bunching up the frayed end and physically

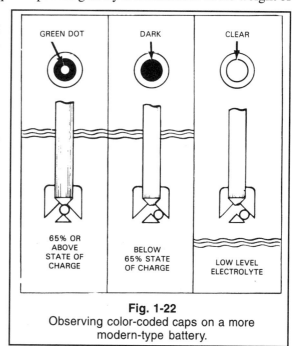

GREEN DOT   DARK   CLEAR

65% OR
ABOVE
STATE OF
CHARGE

BELOW
65% STATE
OF CHARGE

LOW LEVEL
ELECTROLYTE

Fig. 1-22
Observing color-coded caps on a more
modern-type battery.

momentarily joining it to the rest of the battery cable. Soldering serves a two-fold purpose, and one precaution must be followed when soldering. Soldering mechanically secures a junction, which most often consists of two or more metallic parts or surfaces. Secondly, it creates an electrical path through which current can flow virtually unimpeded. The main precaution with soldering, other than the safety factor of not burning yourself, is not to damage any electrical components, such

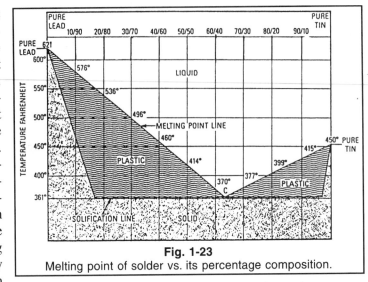

Fig. 1-23
Melting point of solder vs. its percentage composition.

as transistors, etc., you are soldering. The melting point of solder exceeds what most semiconductors can safely withstand. Therefore, by this necessity alone, solder has to melt and solidify quickly, which it certainly does.

Soldering, most often in car repair, is the process of heating a joint composed of metal wires. After heating the wires, you apply solder until it melts. After melting it flows all over the joint, completely covering it. Then it very quickly drops in temperature below its melting point and solidifies. What results is a physically strong, durable, and permanent bond, which has very low resistance. Solder melts best at a specific temperature for the traditional 63 percent to 37 percent ratio of tin to lead (see Fig. 1-23). This melting point at the bottom of that "V" in the graph is solder's minimal melting, or eutectic, point. Eutectic is merely another way of saying solder. There is a proper technique to soldering. Touch the hot soldering iron to one side of the bare wire bundle and apply solder to the other side. If the junction becomes hot enough to melt the solder, it then flows. Allow at least 20 seconds for this to occur. Use a gas-based torch if you can't obtain enough

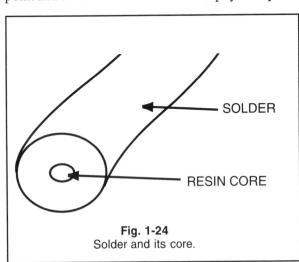

Fig. 1-24
Solder and its core.

heat, even with a large 200-Watt soldering iron. Solders have cores, or are hollow. Figure 1-24 shows the most common type of resin-core solder.

The Concept of Charge and Capacitance

This is the only other formula you will ever use very often in automobile electronics, unless you really delve deeply into its inner-workings. Equation 1-2 relates charge to voltage and capacitance as follows:

Charge in Coulombs = Capacitance in Farads x Voltage in Volts

Stated differently: $Q = C \times V$

or: 1 Coulomb = 1 Farad x 1 Volt

You describe charge in Coulombs. Likewise, you describe capacitance in Farads. This is a quantity of capacity capable of holding a charge. Lastly, as you already know, you describe voltage in volts. Since capacitance represents the ability to hold a charge, what is charge? Charge is, quite simply, just a quantity or packet of potential energy (energy at rest) waiting to become kinetic energy (energy set into motion). Your car's battery is a source of potential energy. When you start your car, your battery is called upon to supply a tremendous amount of kinetic energy in a short burst of time. This potential-to-kinetic energy transformation is what turns over your engine.

A capacitor is an energy reservoir. Therefore, the larger value a capacitor has, the more charge it can hold. In this field we have been working with volts, Ohms and Amps. But a 1 Farad is too unwieldy to work with so we express charges usually in milli-Farads, sometimes even mF which is 1/1,000,000 of a Farad. A 1 Farad capacitor would be about half the size of a VW car, making it impractical for use in any vehicle, except an all-electric car.

What Controls Capacitance?

By remembering our charge-to-capacitance relationship, you can now think of a wire carrying charge to a capacitor or a battery (both energy reservoirs.) The capacitor consists of two plates separated by an insulator. The larger the plates, and the closer they are to each other, the greater the capacitance. There is also a second more practical event which occurs: You now have a better path for the transport of current to charge something which needs a quick supply of energy. Figure 1-25 demonstrates the plate arrangement while Figure 1-26 shows this charging phenomenon.

But unlike a plain resistor, a capacitor acts differently. Figure 1-26 shows that as you close the switch, the current in the circuit with this capacitor maximizes (peaks). Then it steadily "drains off" or decreases as the circuit stabilizes. This is important since in the trouble-shooting procedures section of this book you'll use an instrument called a DSO or (Digital Storage Oscilloscope). This captures fast-fleeting occurrences like this. Once captured and displayed on the CRT, you call them waveforms. This gives you a snapshot, much like a Polaroid camera does. There is more about recognizing and interpreting waveforms in future chapters. This is both an art and a science, and you'll get good at this, trust me. This is because this book presents so many of them.

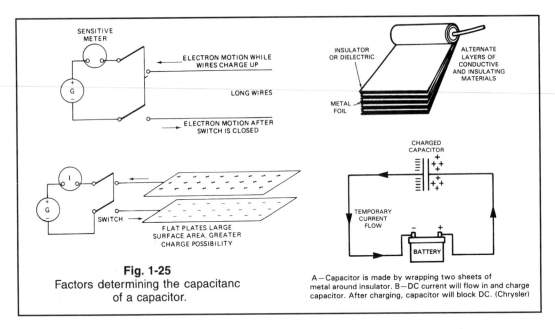

Fig. 1-25
Factors determining the capacitanc
of a capacitor.

A—Capacitor is made by wrapping two sheets of
metal around insulator. B—DC current will flow in and charge
capacitor. After charging, capacitor will block DC. (Chrysler)

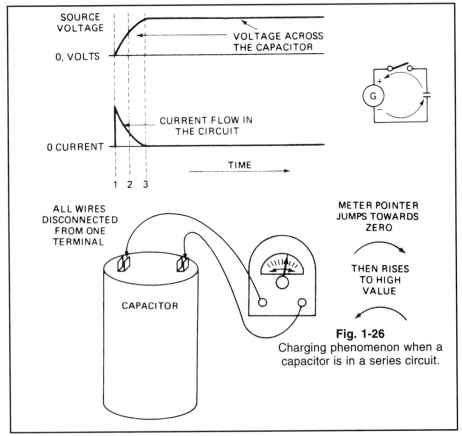

Fig. 1-26
Charging phenomenon when a
capacitor is in a series circuit.

Inductance

This is the opposite of capacitance. Greatly simplified, whenever you cut a wire with a magnetic field, you induce a field of voltage. The only place where we are concerned about this is in the only place where there is AC in your car, in the generator. But now let's see about AC since we now understand DC. The best place to start is with applications of AC.

Alternating Current (AC)

As you recall, AC is the kind of current flow which changes direction. Therefore, this is its main distinguishing trait from DC. Direct current (DC) flows in just one direction from a more positive to less positive voltage source (which is often 0 volts or ground). You may also recall that when you pass a wire through a magnetic field, you induce a voltage in that wire. You can think of this as transforming magnetic energy into electrical energy. The stronger the magnetic field, the higher the voltage. Furthermore, the higher the speed, the higher the voltage.

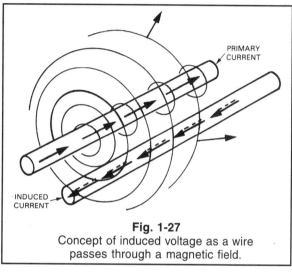

PRIMARY CURRENT

INDUCED CURRENT

Fig. 1-27
Concept of induced voltage as a wire passes through a magnetic field.

But we have said nothing thus far about the direction of this generated voltage. This depends on two circumstances. First, it is the direction in which the wire moves through the magnetic field. Second, it is the direction of the magnetic field itself. Figure 1-27 shows a piece of wire moving through a magnetic field. First, the moving through this field is mechanical force since something has to generate a force behind this wire for it to move. Observe the lower half of Fig. 1-27. Note that in position 1 (the first arrow on the right) the wire has yet to enter this field; however, in positions 2 through 4 it is within this field. In position 5 it has exited the field. A clear pattern emerges in which the voltage increases as it enters the magnetic field and remains unchanged until it exits the field. But in your automobile, you only have such an arrangement in theory.

In practice, you have a rotating coil (Fig. 1-28) which is a generator rotating in a magnetic field. Think about it, a wire going in and out of a magnetic field versus it rotating. The rotating scheme is by far more convenient, as we shall soon see. You can characterize these lines of force by using the right-hand rule (Figs. 1-28 and 1-29). What you have is a wire (or actually a bundle of wires) rotating. If you plot this as a graph of voltage versus time, you'd obtain Fig. 1-30. This constant and repeating up-and-down motion is a sine wave. And each time the voltage crosses the 0V line, its direction of current flow abruptly changes. In a car you experience this with a rotating belt-driven shaft. This mechanically moves these wires (coil) through a generator, or generators A and B in this case (Fig. 1-31). This is typical of what occurs in your car's alternator.

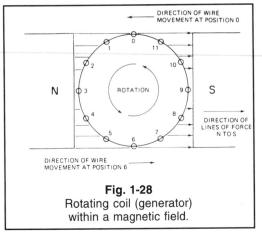

Fig. 1-28
Rotating coil (generator)
within a magnetic field.

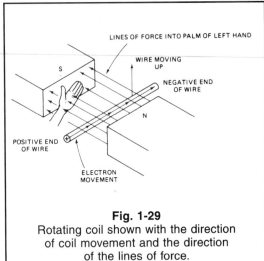

Fig. 1-29
Rotating coil shown with the direction
of coil movement and the direction
of the lines of force.

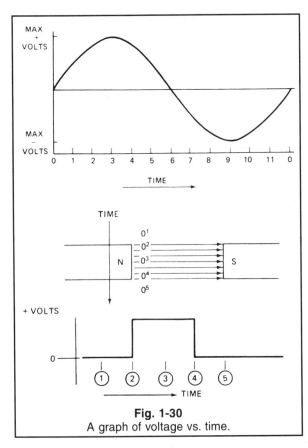

Fig. 1-30
A graph of voltage vs. time.

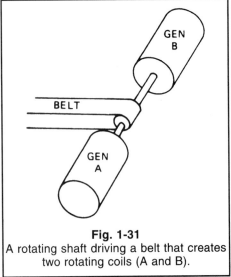

Fig. 1-31
A rotating shaft driving a belt that creates
two rotating coils (A and B).

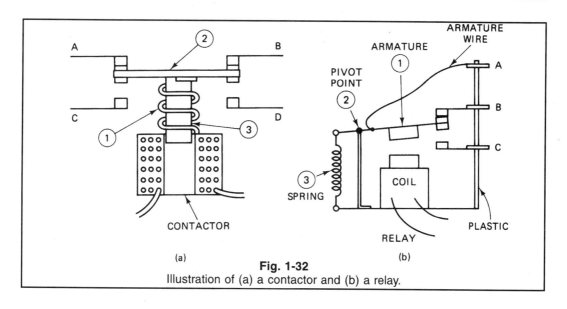

Fig. 1-32
Illustration of (a) a contactor and (b) a relay.

Practical Examples of Troubleshooting with Knowledge Obtained Thus Far

We have seen what a switch can do since it allows you to close and open both a series and parallel circuit(s), and combinations thereof. But, as busy people, we can't idly stand by waiting to throw a switch in a circuit within our car's electrical system. We must come up with a way of having a device that is smart enough to do this switching for us while we are concentrating on other matters, such as driving. There are two devices that are ideal for this task. They are the actuator or "contactor" and the relay (see Fig. 1-32). An actuator is a control device that *works only on AC*. The only place you will ever encounter AC in your car is in circuits associated with your car's generator/alternator.

Referring to Fig. 1-32(a), in its de-energized state, the spring (1) holds the bar (2) in the up position. This allows a complete circuit from terminal A through the contactor's bar at terminal B. A terminal is a convenient place to hook a wire, take a measurement, or make a break or closure in an electrical circuit. This would be represented as a close set of contacts in schematic form. When there is AC at the contactor coil, the coil energizes. The magnetic field pulls the armature (3) to the center of the

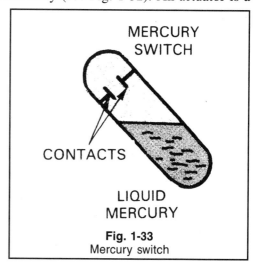

MERCURY SWITCH

CONTACTS

LIQUID MERCURY

Fig. 1-33
Mercury switch

coil. The bar then disengages. It leaves terminals A and B and now makes a new electrical contact with terminals C and D.

Figure 1-32(b) shows a relay. This is quite similar to the actuator in both form and function. The armature (1) rotates around the pivot (2) point. The spring (3) holds the left end of the armature down. This only occurs when you don't energize this coil. When you de-energize this coil, a wire connects terminal A to the armature, and this makes contact between points A and B. We have made an assumption in the preceding descriptions. That is, both of these control devices are composed of metal parts and these are good conductors of electrical current. As good conductors, they provide a minimum of resistance to current flow (back to our water tower example!).

Figure 1-33 shows what you might call a smart switch. The light bulb-shaped object with a pointed tip on its glass envelope, which points to the right, is a mercury switch. True to its name, it is filled about one-tenth full with mercury. Mercury is a liquid substance that easily flows, even when cold. But more importantly, it is a very good conductor. Therefore, even though it is liquid, it can, and very often does, serve as a switch's contact. This mercury switch is in a circuit that might control your air conditioning or heater fan. Another more modern example of a mercury switch is in a theft deterrent device. As a potential thief attempts to possibly lift your car up and tow it away, a mercury switch suddenly experiences this tilting sensation. The mercury then "spills" or assumes a new level. When it does, it allows two new contacts to experience a completed circuit. This drives or turns on an alarm or other warning device such as a very loud horn or light. Lastly, since we have studied capacitors, let's see how to check a mercury switch. Figure 1-34 shows how to do this with a meter (DMM). This indication means it changes states and its properties from an insulator to a conductor. That is typical of a charged versus an uncharged capacitor.

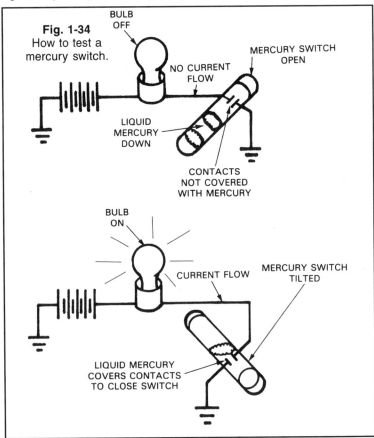

Fig. 1-34 How to test a mercury switch.

BULB OFF

NO CURRENT FLOW

MERCURY SWITCH OPEN

LIQUID MERCURY DOWN

CONTACTS NOT COVERED WITH MERCURY

BULB ON

CURRENT FLOW

MERCURY SWITCH TILTED

LIQUID MERCURY COVERS CONTACTS TO CLOSE SWITCH

Amount of current (mA)	Response
0–0.5	No response
0.5–2	Slight tingling to mild shock; quick withdrawal from body contact
2–10	Mild to heavy shock; muscular tightening
10–50	Painful shock; cannot let go
50–100	Severe shock; breathing difficulties
100–200	Heart convulsion; death
Over 200	Severe burns; breathing stops

1 mA = 0.001 ampere.

Table 1-1
Resistance of the human body and shock hazards per doses of current.

Safety while Working on Your Car

The greatest concern for safety is to avoid receiving a shock. But just what is a shock? It is the passing of electrical current flow through your body. This is largely a variable. This is because most of the resistance occurs within the skin of your hand and fingers. A person with thick calluses has a much higher resistance than a person who has thinner skin and does virtually no manual labor. A typical resistance is 1 MW or 1 MegOhms (1 million Ohms). This value greatly reduces if your hand is wet and/or sweating. If you use Ohm's Law, and assume your car battery has a +12 VDC potential, it is a simple task to calculate current flow through your body. To determine current flow, you would set up to solve for **I** or current. The formula for that is:

$$I = E / R \text{ or } I = 12 \text{ VDC} / 1 \text{ MW} = 12 \text{ microAmps or simply } 12 \text{ mA.}$$

This is 12 one-millionths of an Amp. Table 1-1 shows how insignificant this is. But as previously stated, the real problem is when your hands are wet.

AC is a More Serious Concern

An alternate situation is when you wish to charge your car's dead battery. You use a battery charger and these devices run off of AC wall current. This device takes the wall receptacle's 120 VAC and rectifies it (changes AC into DC) and then applies the 12 VDC to your car's temporarily depleted battery. This can prove far more dangerous of a potential hazard. Symbolically, Figure 1-35 shows the potential for shock with the feet of this man terminating into a ground symbol. This is an alternate way of representing ground potential from that in Fig. 1-8. This actually represents an earth ground symbol. This is where your body touches the earth (through your shoe soles, naturally). The newer type wall receptacles have three holes into which the AC cord's prongs fit. Figure 1-36 shows these three holes as H

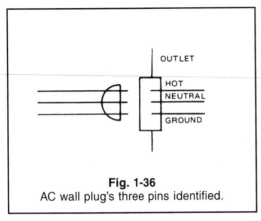

Fig. 1-36
AC wall plug's three pins identified.

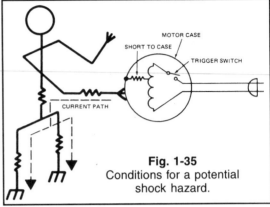

Fig. 1-35
Conditions for a potential
shock hazard.

for hot, N for neutral, and G for ground. But there are a lot of old AC wall receptacles with accommodations for just two prongs.

You can overcome this hazard by buying or easily building a device (Figure 1-37). This has an insulated in-line (series) fuse in the current path of flow to your body. Should you experience a shock which exceeds the current capacity of the fuse, the fuse opens. This is just like a switch because its open position prevents the series flow of DC current. A fuse is the weakest link in an electrical circuit. It is the weakest link because of the manufacturer's purposeful design and selection. It "blows" or opens first. Another potential battery charger hazard is when the case shorts (Figure 1-38). If this occurs, or if you just start to feel the slightest tingle of a shock, repair this charger or throw it away since it is very unsafe to use.

Interconnecting a Car's Electrical Components

This is a crucial indispensable skill that will serve you well throughout your efforts to service and fix your car's electrical system. As we have said, soldering is one way. And when you need extra heat, use a torch (Fig. 1-39). There are some cables typically enshrouding or holding together bundles of finer wires within them (Fig. 1-40). These terminals get crimped or "squeezed" with a wrench, not soldered at all (Fig. 1-40). Note in the upper illustration in Fig. 1-40 that the pliers has a special deformation or curvature to its jaws. This ensures that the metallic coupling or "holder" is properly formed, turned in, and then compressed. The crimped holding/securing metallic fixture in Fig. 1-40 is not by far the only type of clamping device which serves as a termination to a wire in the form of a terminal or similar device in purpose and shape. Figure 1-41 shows numerous different wire connectors. Note, in each case, there is an end that is open. You can insert a wire in here and then solder or tightly crimp it. After you solder it, that exposed solder joint (which is a conductor) might touch and short out something. Therefore, they invented shrink tubing. You can easily cut this plastic tubular shape substance that is pliable, yet rugged. But most importantly, it shrinks when you apply heat to it (see the illustration to the left in Fig. 1-41). After heating, shrunken tubing conforms to the joint and serves as a tight electrically insulating coating.

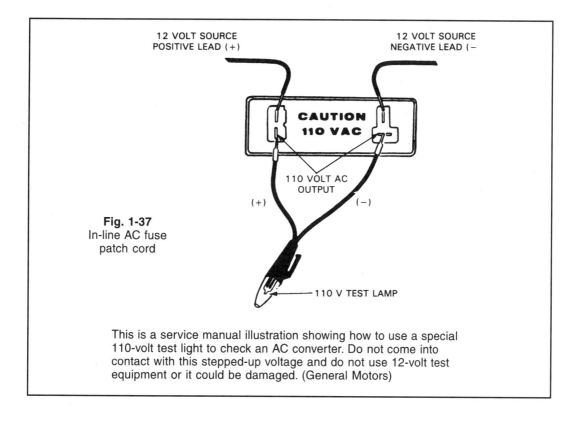

Fig. 1-37
In-line AC fuse
patch cord

This is a service manual illustration showing how to use a special 110-volt test light to check an AC converter. Do not come into contact with this stepped-up voltage and do not use 12-volt test equipment or it could be damaged. (General Motors)

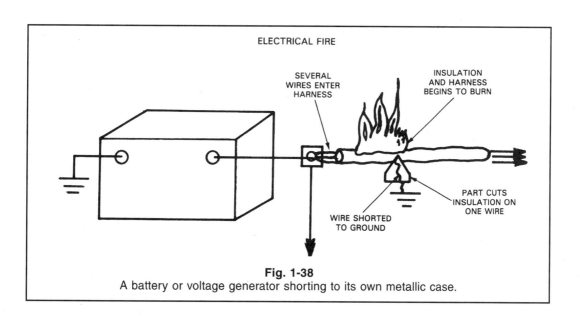

Fig. 1-38
A battery or voltage generator shorting to its own metallic case.

Another approach to using ***shrink tubing*** is as an electrical insulation around a current-carrying conductor. You can do this by following the step-by-step instructions in Fig. 1-42. This technique most aptly applies to multistranded wires, which run to your car's computer. As Murphy's Law would have it, you may find a broken wire within a complex wiring harness. A ***wiring harness,*** see Fig. 1-43(a), is typically lacing, tying, or assembling numerous wires into a neat bundle and the manufacturer routes them to a specific spot to avoid an unnecessarily untidy clutter. Not using a wiring harness would make servicing and identification of a specific wire(s) far more confusing and difficult. We covered fuse links, in the form of an in-line AC fuse, in our safety discussion, but the lower illustration in Fig.1-43(b) shows actual pieces of wire serving as ***fuse links***. This means that a wire has a specific current-carrying capability. Such a wire can serve as a fuse since a fuse ruptures when you exceed its rated maximum current. Obviously, carefully controlling manufacturing tolerances of the ***rupture current*** is crucial. This is the current that explodes and breaks away the wire. There can be no more than a +/– 10 percent variation in rupture current. Otherwise, you'd overstress delicate electronic components.

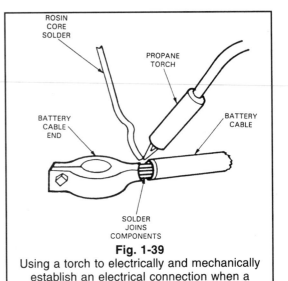

Fig. 1-39
Using a torch to electrically and mechanically establish an electrical connection when a soldering iron will not supply enough heat.

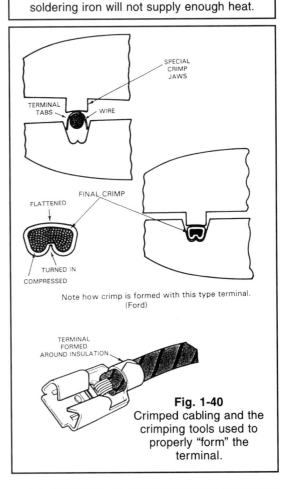

Note how crimp is formed with this type terminal.
(Ford)

Fig. 1-40
Crimped cabling and the crimping tools used to properly "form" the terminal.

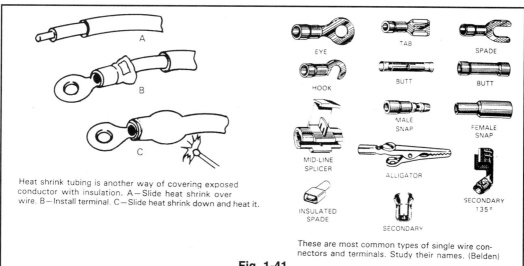

Heat shrink tubing is another way of covering exposed conductor with insulation. A—Slide heat shrink over wire. B—Install terminal. C—Slide heat shrink down and heat it.

EYE

TAB

SPADE

HOOK

BUTT

BUTT

MALE SNAP

FEMALE SNAP

MID-LINE SPLICER

ALLIGATOR

SECONDARY 135°

INSULATED SPADE

SECONDARY

These are most common types of single wire connectors and terminals. Study their names. (Belden)

Fig. 1-41
Various types of wire terminals and connecting devices.

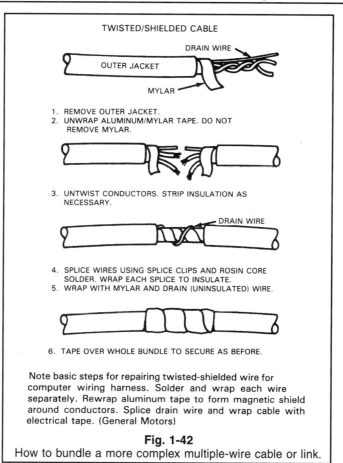

TWISTED/SHIELDED CABLE

DRAIN WIRE

OUTER JACKET

MYLAR

1. REMOVE OUTER JACKET.
2. UNWRAP ALUMINUM/MYLAR TAPE. DO NOT REMOVE MYLAR.

3. UNTWIST CONDUCTORS. STRIP INSULATION AS NECESSARY.

DRAIN WIRE

4. SPLICE WIRES USING SPLICE CLIPS AND ROSIN CORE SOLDER. WRAP EACH SPLICE TO INSULATE.
5. WRAP WITH MYLAR AND DRAIN (UNINSULATED) WIRE.

6. TAPE OVER WHOLE BUNDLE TO SECURE AS BEFORE.

Note basic steps for repairing twisted-shielded wire for computer wiring harness. Solder and wrap each wire separately. Rewrap aluminum tape to form magnetic shield around conductors. Splice drain wire and wrap cable with electrical tape. (General Motors)

Fig. 1-42
How to bundle a more complex multiple-wire cable or link.

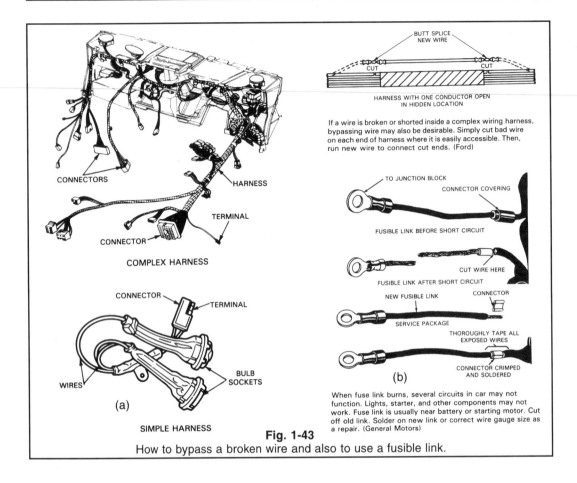

Fig. 1-43
How to bypass a broken wire and also to use a fusible link.

Chapter I Quiz

True or False

1. The author feels strongly that modern cars are far too complex to fix.
2. The water tower analogy helps you visualize what Ohm's Law really means, especially when you are trying to understand electrical units relations to one another. It is also quite useful for when you are making calculations.
3. Almost everything in your car runs off of AC.
4. Your car battery produces both AC and DC.
5. The pie chart with three sectors representing current, resistance, and voltage helps you better remember Ohm's Law.
6. Exponents to even power are called scientific notation.
7. An analog ammeter has a digital display.
8. A car battery's "–" terminal is conventionally termed its ground.

9. Loads are energy restoring devices, while current circulates within your car's electrical circuits.

10. The "loop" appearing symbol within a glass lamp's symbol represents its filament.

Matching

_____ 11. This is the electrical commodity that is common to a series circuit.

_____ 12. This is the electrical commodity that is common to a parallel circuit.

_____ 13. This is the voltage most car batteries have.

_____ 14. These lights on your car consume more current than all others combined.

_____ 15. This quantity seeks the path of least resistance.

_____ 16. This device allows current flow when you close it.

_____ 17. An open switch _____ current flow.

_____ 18. This is the substance in which your battery's internal portion dwells.

_____ 19. A car's well-frayed battery cable results in _____.

_____ 20. What determines the difference in the two types of car batteries?

A. Voltage

B. Blocks

C. Battery drainage

D. The position of the battery's caps.

E. Current

F. Sulfuric acid.

G. A switch.

H. Current

I. 12 Volts

J. Headlights

Multiple Choice

_____ 21. You should apply heat to a soldering joint. If at the end of _____, the solder doesn't melt and flow over the joint, you need to move up to a heavier heat-producing device.

A. 2 seconds

B. 2 minutes

C. 1 hour

D. None of the above.

_____ 22. You have overcharged your car's battery when _____.

A. It generates gas.

B. It vents this gas to the outside world through a vent on your battery.

C. It overheats and then gets quite cold.

D. A and B.

_____ 23. The unit of charge is the _____.

A. The Farad

B. The Weber

C. The Ohm

D. None of the above.

_____ 24. A good car battery should have a hydrometer reading of from 1.265 to 1.299. This represents the following _____.

A. It represents the specific gravity of your car battery's electrode.

B. It visually indicates the specific gravity of the battery's acid to water.

C. It is the desired ratio between the weight (per unit volume) of the battery acid to an equal volume of water.

D. All of the above.

_____ 25. The color-coded caps on more modern car batteries indicate _____.

A. The temperature within your car's battery.

B. The amount of wear on your car's overall electrical system.

C. The specific gravity of the car battery's electrolyte, thus its ability to hold a charge.

D. All of the above.

_____ 26. Soldering has two purposes, these are _____.

A. To mechanically and physically secure the solder joint.

B. To establish a direct path to ground.

C. To create an electrical path, enabling virtually unimpeded current flow.

D. A and C.

_____ 27. The ratio of lead-to-tin in solder most dominantly affects what?

A. How smooth the joint will eventually be.

B. How strong the joint will be in the cold.

C. The melting point of the solder.

D. All of the above.

_____ 28. Inductance is the opposite of what?

A. Resistance

B. Capacitance

C. Voltage

D. Charge

2

The Cooling System:

Getting Acquainted with a DMM

to Troubleshoot Your Cooling System

Your car's cooling system is simple by comparison to most of your other automotive systems. It is, therefore, an ideal place to "as painlessly as possible" start our actual troubleshooting procedures. But before we start troubleshooting, we need to gain a better understanding of how the simplest troubleshooting instrument works. This is the digital multimeter (DMM). So let's start with your cooling system and then move to the DMM discussion in the second half of this chapter.

Figure 2-1 illustrates the main functions and components of a car's cooling system. After becoming familiar with them, you'll investigate how to both diagnose and fix cooling system malfunctions. This discussion assumes:

1. You live in a diverse enough climate to require a car heater.
2. You have an internal combustion engine car less than 30 years old.

This all-too-often overlooked system definitely deserves more respect, appreciation, and, most certainly, better treatment. As examples, when cars change engines to newer or rebuilt ones, less than 10 percent of them have anything done to their radiator and/or cooling system. However, 55 percent of all engine replacements stem from a cooling system/radiator malfunction. Next, the second most common road hazard, only surpassed in frequency by a flat tire, is a radiator problem.

The Radiator's Job

It's the key ingredient in a cooling system whose purpose is to regulate temperature. Most people only think of it as a cooling system, which it is—even by its very name; however, if your car runs too cool (and this can happen), it not only wastes fuel, but causes condensa-

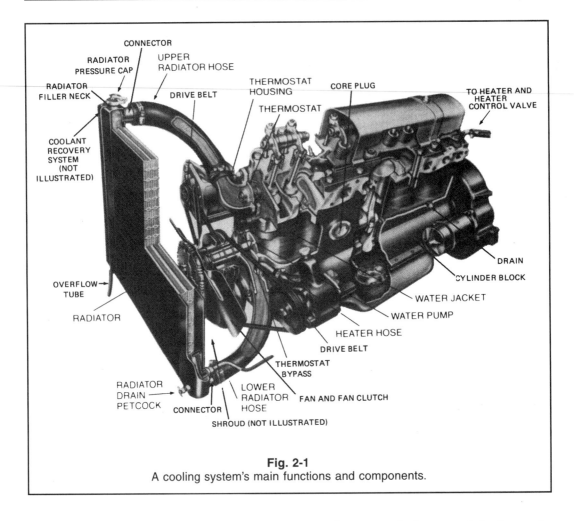

Fig. 2-1
A cooling system's main functions and components.

tion. This promotes abnormally severe piston wear. It also causes sludge to settle in the crankcase. Running your car too hot causes the far more familiar effects of:

1. Seizing of an engine's bearings and pistons.
2. Burning of valves.
3. Scored cylinder walls.

Lesser Well-Known Radiator Facts

These largely go unappreciated and are:

1. Using an abnormal rate of lubricant.
2. Excessive internal engine heat buildup, expanding and contracting the engine's parts beyond acceptable tolerances. This ruptures gasket heads and leads to oil leaks.
3. Some radiators are further burdened by cooling your automatic transmission fluid.

A Radiator's Operating Environment

There is considerable metal-to-metal rubbing within your car's engine. This produces friction, and, in turn, heat! Your car's cooling system removes (or exchanges) this heat to the outside cooler ambient air. This is a basic, very straightforward and easy-to-understand example of *heat transfer*. There are three types of heat transfer. The first is radiation, in which heat from a hot surface (typically a metal) mixes with the outside cooler air. The second is conduction, in which two thermally conductive surfaces or masses (metals) touch and a compromise intermediate temperature results in both metals. A heat sink on a power transistor is a prime example of this. The last is convection in which a device, such as a fan, vigorously circulates surface-heated air. This cools by mixing this heat with its gaseous surroundings (most often cooler air). You can visualize convection as just radiation being given a boost by a fan, which sits behind the radiator. Naturally, it greatly helps to evacuate the hot surface's heat to the outside cooler ambient air.

A radiator advantageously exploits all three of these methods of heat transfer. However, air alone is an inadequate cooling medium; therefore, a radiator uses an internal fluid. This fluid circulates at a rate of from 2,000 to 7,000, or more gallons per hour—actually twice that fast in a Porsche Carrera! Water alone boils away too quickly to serve as an effective coolant. Most often, this fluid is 50 percent water with 50 percent ethylene glycol. This mixture boils at a temperature approximately 25 degrees higher than water alone (212°F + 25°F = 237°F). This mixture satisfies the radiator requirements of most modern higher compression engines.

The Deterioration Mechanism and Constraints on Modern Cars

Temperatures within modern high-compression engines reach 220°F or higher. There is a triple whammy, though, lurking here. First, this obvious "added heat overstressing" constraint combines with an ever decreasing amount of space under the hood. Secondly, heat causes antifreeze breakdown. This lowers the coolant's pH making it more acidic and prone to eat metal (engine parts). Your poor radiator is the victim and has to surrender its previously occupied space to devices like air conditioners, control devices, brake boosters, etc. Nonetheless, despite all these detracting factors, your radiator has to perform far better than its predecessors—yet within these greatly reduced spaces, and at higher normal internal operating temperatures than cars of past generations.

But How Do You Know?

The most obvious results of antifreeze breakdown—and the subsequent high acidity and metal-eating properties of this situation—are leaky seals, cracks in metallic parts and other problems which cause antifreeze to seep into your car's motor oil. Realizing this vital "need to know", Blackstone Labs (http://www.blackstone-labs.net 219-744-2380), sells a kit for $18.50

EXAMPLE OIL REPORT

P.O. NUMBER: Verbal, Kevin
CODE: 001

UNIT NUMBER: 94 SAAB 9000
REPORT DATE: 08/04/98

CLIENT

CONTACT: Kevin Banks
NAME: GUILFORD SAAB
ADDRESS: 1800 Boston Post Rd
Guilford, CT 06437

LAB NUMBER: B13700
PHONE: (203) 453-0180
FAX: (203) 453-5920

UNIT

EQUIPMENT MAKE: Saab
EQUIPMENT MODEL: 9000, 4-cyl 2.3L
FUEL TYPE: Gasoline (Unleaded)

OIL USE INTERVAL:
OIL TYPE & GRADE: 10W30 (Gas)
MAKE-UP OIL ADDED:

COMMENTS

KEVIN: All wear you can get from this type engine was abnormal, likely showing a catastrophic failure. The sample submitted was close enough to being a solid that physical testing was impossible. The low flash point may show 6.7% of the sample was gas. We doubt any anti-freeze exists. Silicon may be from failing parts. This oil has been run against hot parts and has been added to extensively. R1009394

ELEMENTS IN PARTS PER MILLION

EQUIPMENT HOURS/MILES		UNIT / LOCATION AVERAGES								UNIVERSAL AVERAGES
SAMPLE DATE	08/03/98									
ALUMINUM	40	0								8
CHROMIUM	24	0								4
IRON	754	0								135
COPPER	90	0								29
LEAD	393	0								105
TIN	31	0								6
MOLYBDENUM	18	0								16
NICKEL	7	0								1
MANGANESE	15	0								2
SILVER	0	0								0
TITANIUM	0	0								0
VANADIUM	0	0								0
BORON	29	0								27
SILICON	37	0								10
SODIUM	125	0								189
CALCIUM	1785	0								1330
MAGNESIUM	546	0								421
PHOSPHORUS	892	0								992
ZINC	1348	0								1175
BARIUM	4	0								20

PROPERTIES

TEST	cST VISCOSITY @ 40°C	SUS VISCOSITY @ 100°F	VISCOSITY INDEX	cST VISCOSITY @ 100°C	SUS VISCOSITY @ 210°F	FLASHPOINT IN °F	FUEL %	ANTIFREEZE %	WATER %	INSOLUBLES %
VALUES SHOULD BE						59–72	>365	<2.0	0	<0.05	<1.1
TEST VALUES WERE						–	230	6.7	0	0.0	35.0

© COPYRIGHT BLACKSTONE LABORATORIES 1996

4929 S. LAFAYETTE STREET
FORT WAYNE, IN 46806 (219) 744-2380

LIABILITY LIMITED TO COST OF SAMPLE ANALYSIS

Fig. 2-2
A wear and general status "engine" health report.

which you give the person changing your oil. A portion of the old oil fills a bottle and you send it back to them. The postage is $1.47. Blackstone runs an analysis on the oil and lets you know if any antifreeze has seeped into places where it has no business being. Figure 2-2 is a typical report.

A Word of Caution

There seem to be many heated controversies concerning exactly what constitutes a "safe" coolant. If you err, err on the side of caution by avoiding any practice or substance that your owner's manual does not specifically call out. Also, as preventative measures, check your belts and hoses very often, and if your radiator springs a leak, have it welded. Do not believe in these TV-advertised leak fix products, which are largely ineffective.

Brass and copper usually "mostly" comprise a radiator. Therefore, if you use the wrong coolant it causes flaking or microscopic droppings (particles). This

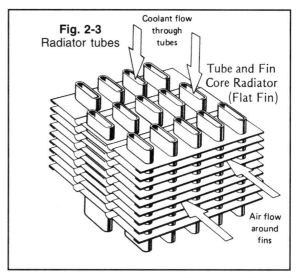

Fig. 2-3 Radiator tubes

Coolant flow through tubes

Tube and Fin Core Radiator (Flat Fin)

Air flow around fins

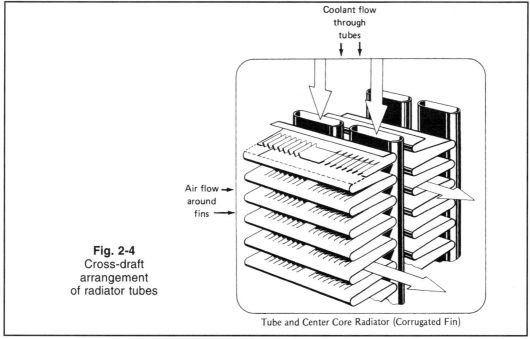

Coolant flow through tubes

Air flow around fins

Fig. 2-4 Cross-draft arrangement of radiator tubes

Tube and Center Core Radiator (Corrugated Fin)

clogs the circulation of the cooling system. To alleviate some possible confusion, most radiators corrode. Corrosion is metal deterioration which is either chemical and/or electrochemical in nature.

Corrosion continues until the corrosive material is removed from the metals. That's why you should flush your radiator regularly. In contrast, rust is a special case of corrosion which occurs between iron and steel. The actual "constriction" mechanism results when these microscopic, and other flakes, clog your radiator tubes (see Figure 2-3). A tube is a "pocket" which holds the coolant. A typical arrangement occurs when these tubes are in line with the direction of your radiator blade's induced air flow across it (see Figure 2-4).

The Nitty-Gritty or Getting Down and Dirty

To more fully appreciate the constraints of this operating environment, 2800°F (twice the temperature to melt iron!) is not at all uncommon within your modern internal combustion engine. The tremendous power resulting from a combustible mixture of gas and oxygen exploding within your cylinder walls drives your pistons up and down through their connecting rods, etc. But if that were not bad enough, only one-third of this created energy goes to pushing the pistons up and down. One-third goes out the exhaust as hot air, and your radiator has to handle the last third. More dramatically, one of these thirds of internally combustible-created energy is enough to heat a five-room house to 70°F when it is 0° outside. WOW! Maybe now you realize how important and tough of a job your car's radiator does.

Common Radiator Failure Modes

The worst culprit is an obstruction in the circulating path within the cooling system. Following our established description of the cooling system, let's now stress its flow "mechanism" throughout all your engine parts—including your heater. Obviously, if something either impedes (slows down) or "pinches off" (completely obstructs) the flow, then either insufficient, or worse yet, no heat exchange and "venting" occur.

Let's Thank Newton

Newton's Law of Thermodynamics states that the ratio of two heat-possessing masses (and air is a mass) has emissivity (the ability to give off heat from a surface) in proportion to the fourth power. In easier-to-understand practical terms, you are driving in Arizona and it's 100°F. Your car's radiator is 200°F. Therefore, if you take the fourth power of 200 and divide it by the fourth power of 100, you have 16. And 16 times is how much more thermal energy the radiator has to emit than the outside ambient air possesses.

As materials heat (particularly metals), they almost always expand, and a radiator is no exception to this "rule of thumb." A positive temperature coefficient describes this expansion phenomenon in relation to heat. In your case, the radiator virtually becomes a self-contained pressure cooker. Therefore, it requires a preset pressure cap. This pressure also unfortunately raises the coolant's operating temperature. Elevated pressure is beneficial in some ways, though, since it drives fluid throughout the entire cooling system to perform its task of heat transfer to the outside.

The Cooling System's Anatomy

If you subdivide your cooling system into a detailed listing, you arrive at the following 16 constituent components:

1) Hoses and connectors
2) Fans and shrouds (covers)
3) Drive belts
4) Water (coolant) pumps
5) Cylinder block and head
6) Core (freeze) plugs
7) Thermostat
8) Thermostat coolant bypass

9) Radiator filler neck
10) Surge tank
11) Overflow tube
12) Radiator pressure cap
13) The radiator itself
14) The coolant recovery system
15) The heater
16) The heater's control valve

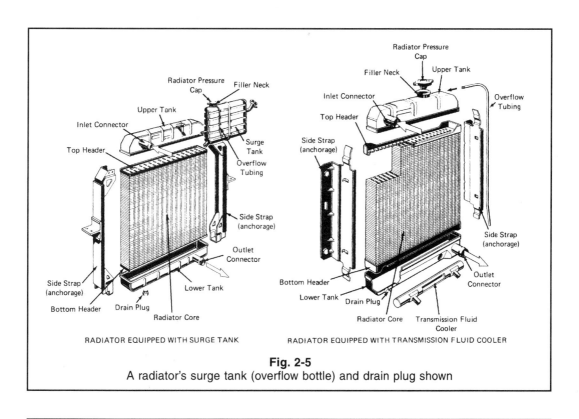

Fig. 2-5
A radiator's surge tank (overflow bottle) and drain plug shown

RADIATOR EQUIPPED WITH SURGE TANK

RADIATOR EQUIPPED WITH TRANSMISSION FLUID COOLER

Let's examine these cooling system constituent components, both separately and together, to gain a feel for how they interact with one another.

The Hose and Connectors

There are two kinds of hoses:

1. Elastometric a (stretchable) flexible and pliable material.
2. A single piece of molded rubber.

The hose naturally connects the radiator to the engine. It therefore serves as a conduit (passageway) for coolant fluid. The bottom, or outlet hose, serves as the output tank for the radiator (see Figure 2-5). The cooled coolant moves through this connector to the water pump and into your engine. Hot coolant returns from the engine to be perpetually recycled.

Be forewarned that the outlet hose often has a spiral interior wire. This gives it form, and rigidity which prevents it from collapsing under a vacuum. Preventing a deformation of shape or an outright collapsing ensures a constant pressure flow of coolant. However, this spiral wire can rust. These flakes are yet another source of microscopic droppings, which by themselves, or cumulatively and collectively, contribute to your radiator's clogging. Since you can't see this from the outside, you should regularly check your inlet hose. The hose securing mechanisms are clamps of two varieties: 1) spring-screw types and 2) wire clamps. The former are better.

Fans

There are several varieties of fans: thermostatic, centrifugal, fluid coupling, and flexible blade. Most types are fixed rigid blades that the water pump pulley drives directly through either cap screws or studs. Since your engine's speed determines the rate of rotation on this fan, it is understandable that in stop-and-go traffic, or worse yet, grid-locked traffic, the heat venting effectiveness is greatly compromised by this slowly revolving fan. This is how radiator overheating occurs.

Blades are crucial, since manufacturers use varying numbers of blades and design them to perform optimally at a specific pitch. Therefore, once bent, this sophisticated balancing scheme no longer exists, and, unfortunately, blade replacement is your only alternative. The flexible blade fans mentioned earlier flex or "flatten out" at higher engine speeds to "cut the wind" more easily to help save fuel.

Variable-pitch (flexible) fans use two types of clutches. The first, the centrifugal clutch, runs in proportion to your engine's speed. Once you reach a sufficient speed, it disengages (releases) the fan. The second, the thermostatically controlled clutch, uses temperature sensing (or measurement) to determine when your car disengages its fan clutch. Once you reach an adequate speed to have the fan disengage, your car's front-end design allows "ram air" to directly hit your radiator. These miniature "spoilers" are called shrouds.

Keep things in perspective, regardless of your fan blade design, because there are limitations. Specifically, when you reach about 65 mph (or faster), the fan blades only negligibly

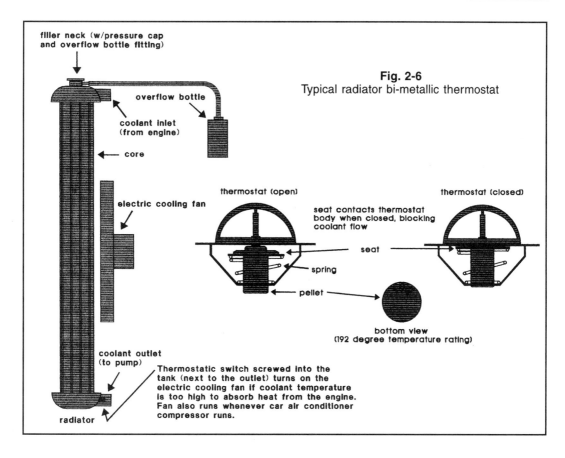

filler neck (w/pressure cap and overflow bottle fitting)

overflow bottle

coolant inlet (from engine)

core

Fig. 2-6
Typical radiator bi-metallic thermostat

electric cooling fan

thermostat (open)

thermostat (closed)

seat contacts thermostat body when closed, blocking coolant flow

seat

spring

pellet

bottom view
(192 degree temperature rating)

coolant outlet (to pump)

Thermostatic switch screwed into the tank (next to the outlet) turns on the electric cooling fan if coolant temperature is too high to absorb heat from the engine. Fan also runs whenever car air conditioner compressor runs.

radiator

help cool your radiation. At these speeds, the sheer in-rush of the oncoming air overwhelms any possible venting effect your fan blades can render.

The Drive Belt

The drive belt(s) turn the fan and the water pump for the cooling system, see Fig. 2-6. They also turn the alternator for a replenishing supply of current flow to your car's lead-acid storage battery. There can be from one to four belts directly driven by the crankshaft's pulley. These pulley-driven belts turn at a rate which is directly proportional to your engine's speed.

The Water (Coolant) Pump—The System's "Heartbeat"

This device usually has a die-cast aluminum or cast-iron housing. It circulates coolant to the heat fins and tubes, from where the heat vents to the outside ambient air. In a typical car, a water jacket surrounds the combustion chamber and associated components, i.e., the block, head, and intake manifold.

Core "Freeze" Plugs

These are plugs made of a thinner softer aluminum than the engine block and are the equivalent of a fuse in an electrical circuit. But instead of too much current causing an open circuit, too much heat buildup causes them to expand and just pop out. They are often vulnerable to leaks and corrosion.

Thermostat

Most thermostats are solid expansion "pill" types, see Figure 2-6. A thermostat resides between the front of the engine and the radiator's top or inlet hose. Upon start-up, your engine is cold, and the thermostat does not initially allow coolant circulation. After the

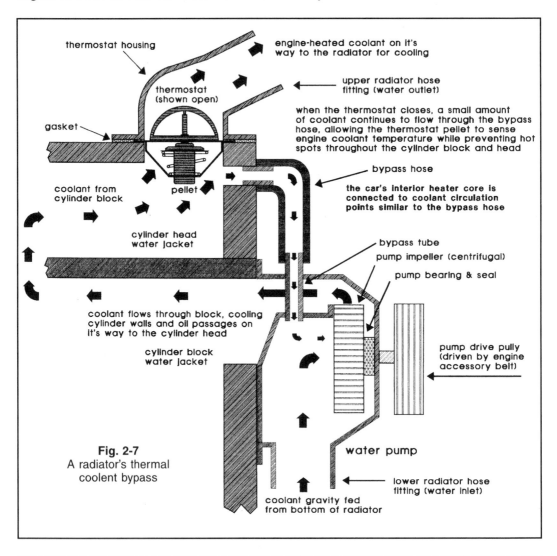

Fig. 2-7
A radiator's thermal coolent bypass

thermostat housing

engine-heated coolant on it's way to the radiator for cooling

thermostat (shown open)

upper radiator hose fitting (water outlet)

when the thermostat closes, a small amount of coolant continues to flow through the bypass hose, allowing the thermostat pellet to sense engine coolant temperature while preventing hot spots throughout the cylinder block and head

gasket

bypass hose

coolant from cylinder block

pellet

the car's interior heater core is connected to coolant circulation points similar to the bypass hose

cylinder head water jacket

bypass tube
pump impeller (centrifugal)
pump bearing & seal

coolant flows through block, cooling cylinder walls and oil passages on it's way to the cylinder head

cylinder block water jacket

pump drive pully (driven by engine accessory belt)

water pump

coolant gravity fed from bottom of radiator

lower radiator hose fitting (water inlet)

engine reaches a temperature—varying from 160 to 220°F—the thermostat's temperature-sensitive valve opens, allowing coolant to start cooling the now "warmed-up" engine. As previously stated, running your engine too cool causes problems can cause moisture buildup in the combustion chamber.

Thermostat Coolant Bypass

This thermostat function allows coolant to reach the engine's "hot spots" during the warm-up period. The coolant flows past the engine's water jacket, but not yet flowing entirely throughout the entire cooling system. Once the engine reaches its necessary "trip" temperature, the thermostat's coolant bypass has completed its task and unrestricted cooling system flow freely occurs (see Figure 2-7).

The Radiator Filler Neck

This is the connector between the pressure cap and the top of the radiator. It has two functions:
1. It serves as a funnel for coolant which pours into the radiator.
2. It properly positions or "seats" the pressure cap. There is usually a small metal nipple on its side. This connects to the overflow tubing.

Surge Tank

If your car's cooling system does not have a filler neck, a surge tank, located next to the radiator, is an alternate avenue by which coolant pours into the cooling system.

Overflow Tubing

This tube may be formed out of copper, steel, or a flexible compound. It attaches to the side of the filler neck. As your cooling system starts to overheat, it causes the coolant to boil. This device is another automotive equivalent of an electronic fuse, since it acts as a relief pressure valve. The coolant boils and overflows to the ground, preventing "radiator bulging" from excessive pressure buildup.

The Radiator Pressure Cap

Many modern cars require a pressure cap for their water pumps to work properly. This either closes or opens the filler neck. Its added function is to pressurize your cooling system. As previously stated, your coolant can reach 237°F or higher. This assumes you are at sea level, since, to be very technically correct, atmospheric pressure raises 3°F or 1.7°C per kg/cm. To better visualize what is happening, the greater the difference between the temperature

Fig. 2-8 Radiator Pressure Caps: (Left) Safety Lever Vent Type; (Right) Coolant Recovery System Type.

of the coolant and the outside ambient air, the faster the heat dissipates. This is Newton's Law of Thermodynamics in a practical, more understandable example.

The pressure cap consists of a pressure valve and a vacuum release valve (see Figure 2-8). There are two kinds of release valves:

1. A closed or constant pressure valve.
2. An open pressure or vent type.

Based on laws of physics concerning compressing a gas, the cooling system's ability to more effectively cool increases in a pressurized closed system. That is why your engine cools substantially without the water ever reaching 212°F, or the boiling temperature of water. But this does not imply that water alone exists as your coolant.

The Fly in the Radiator's Ointment

The quality of the water you mix in your radiator is very important. Distilled or deionized water should accompany concentrated antifreeze. Most city tap water contains both calcium and magnesium, and both react with rust inhibitors in the radiator with a resulting crystallization and precipitates (fall out as) solids. Also, large amounts of chlorides and sulfates, which are common to treated tap water, corrode and pit metals.

Special Antifreezes

The TORO Parts Department has an antifreeze which Puegeot, the French carmaker, recommends. This is P/N 93-7213 and its claim to fame is that it is a nonfoaming type, has a pH balance to maximize heat transfer, and is ideal for aluminum cylinder heads and engine block cars. Valvoline/Zerex has an antifreeze used by BMW which is environmentally friendly and both nitrate and sulfate free, contact 1-800-832-6825.

Finally, the Radiator!

The radiator and the water pump are the two most vital components in your car's cooling system. As you can no doubt logically deduce, for your cooling system to properly function, the radiator must vent heat to the outside very efficiently. However, the radiator's core has to first receive this hot liquid mixture of usually 50 percent antifreeze and 50 percent water. This is what the water pump does. It is called a water pump, despite it actually pumping a water and antifreeze mixture. The water pump circulates water from around the engine jacket, to cool the block, and on to the radiator. There the air cools it, and then it is pumped to the engine block again. This cooling and heat exchange cycle repeats itself or constantly recirculates. Naturally each time it comes from the radiator it is cooler, since this is what your cooling system is all about.

Getting Acquainted With Your DMM

A DMM *Tutorial With Applications*

A modern digital multimeter (DMM) is a handy device that can eliminate a lot of problems around the house and in your car. Modern DMMs do much more than their counterparts of just five years ago. As an example, they now have both visual and audible continuity, peak hold, level detection, conductance, diode checking, and even counting with a frequency counter built in some more modern DMMs. Some even come with a temperature measurement capability option as well, as we will soon discover.

Fig. 2-9
A modern, sophisticated DMM
(Courtesy Fluke Corp.)

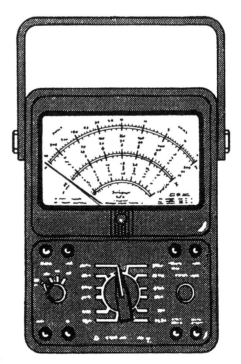

Fig. 2-10
An older analog multimeter

Modern DMMs (Digital Multimeters) come in a variety of shapes and forms, but most look like the one in Fig. 2-9. These replaced and overtook in technology and ease of use the older Simpson 260 type multimeters (Fig. 2-10) with which you may be familiar. This section is a tutorial on DMMs and begins quite simply with the basics. Let's begin by examining the DMM's operation and make a few rather simple measurements, such as determining battery voltage and doorbell continuity in a troubleshooting application. We conclude by making temperature measurements and examining the electrical system in your house or place of business.

Getting Started

We'll begin by using the Fluke model 88 DMM. Before delving into the modern DMM, let's first realize that its predecessor, the analog multimeter, had a d'Arsonval movement as its heart, a swinging needle (Fig. 2-11). The resistors are selected by range and function switches to scale the input for the appropriate reading. Input voltages are reduced by series resistors to magnitudes where the meter movement can handle it, or else you will rapidly "ping" the needle. Shunt resistors develop a voltage proportional to the current which is read by the movement and shown as amps. If you wanted to measure ohms, current was generated by a battery inside the meter and sent through a known resistance. The range switch selected a resistor which would provide a sufficient voltage drop for good accuracy.

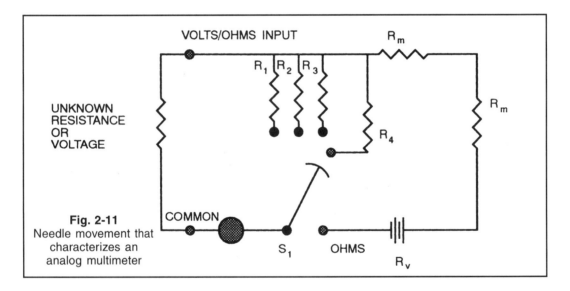

Fig. 2-11
Needle movement that characterizes an analog multimeter

Modern DMMs, like their ancient counterpart, the analog multimeter, still measure resistance, current and voltage, but they do it in a much different fashion. Rather than send the converted voltage down to the meter movement, it is routed to an A/D converter (Fig. 2-12). More factors affect this instrument's accuracy than with an analog multimeter. The accuracy is affected by the quality of the resistors chosen, the resistor network, the oscillator driving the clock, and the basic accuracy of the A/D converter.

Let's examine the more important blocks of the DMM's block diagram. The signal conditioner converts the incoming voltage to a DC voltage and scales it down (reduces) to a level the A/D converter can accept. Next, either an averaging or an rms (Root Mean Square) converter transforms the AC current to a voltage across a current shunt. A current shunt is a precise current diverter, meaning it allows only something like 1/10 or 1/100 or even 1/1,000 the current to pass through the sensing circuitry. If the whole current were to pass through it, it could easily destroy the DMM's delicate input circuitry.

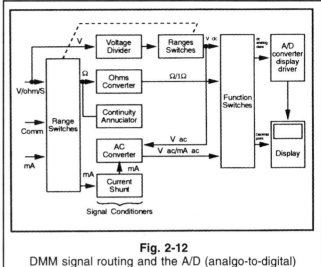

Fig. 2-12
DMM signal routing and the A/D (analgo-to-digital)

The real heart of a DMM is the A/D converter. One popular method of conversion is the dual slope integrating A/D converter. This method measures the time required to discharge a capacitor from a known voltage. The discharge time is proportional to the original analog signal and is measured by the clock pulses from the microprocessor (Fig. 2-13). The interaction of the capacitor with the op amp ensures that the discharge rate is always the same (is linear), regardless of the voltage to which it has been charged. A normal charge/discharge curve in an RC network bows because of the exponential nature of the charge/discharge; however, the op amp provides a constant current source which makes the charge/discharge curve very linear.

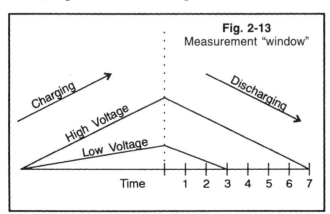

Fig. 2-13
Measurement "window"

The AC converter changes the AC voltage into DC voltage. In many DMMs the accuracy is affected by crest factor (to be explained shortly), frequency response, and bandwidth. The shape of the waveform (Fig. 2-14) is as important as the frequency. The averaging-type AC converter is designed to produce a DC voltage proportional to the average over a period of time.

Before learning how to make actual measurements, let's examine AC and DC coupling. An AC coupled DMM measures the AC component of a waveform independently of any DC component. The total rms value of both AC and DC though can be obtained by the root sum squared calculation after measuring the DC and AC components:

$$[AC + DC] = [(AC)^2 + (DC)^2]^2$$

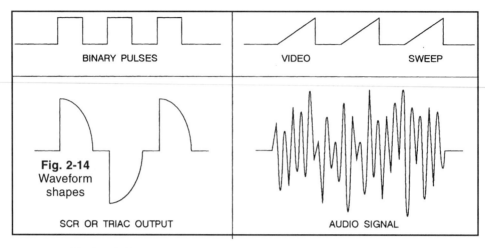

BINARY PULSES   VIDEO   SWEEP

Fig. 2-14
Waveform
shapes

SCR OR TRIAC OUTPUT   AUDIO SIGNAL

Using a DMM

For safety's sake, the test leads and function switch must be properly placed. This may seem obvious, but failure to do so can result in a mishap.

Measuring Voltage and Resistance

When the DMM is connected in parallel across a component, you can measure its voltage (Fig. 2-15a). The same DMM can make a resistance measurement (Fig. 2-15b). When semiconductors are found in the circuit you are testing, it can be difficult to read the value of a component connected to these devices. To get around this, the device must be unsoldered. This is obviously a hassle. So, some more modern DMMs have a very low test voltage. The voltage in the Fluke model 88 is 0.3V, which won't affect the semiconductor junction to make it conduct. But while not affecting this P-N junction, it does allow a resistance measurement to be made.

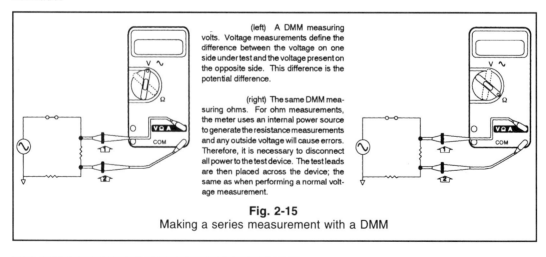

(left) A DMM measuring volts. Voltage measurements define the difference between the voltage on one side under test and the voltage present on the opposite side. This difference is the potential difference.

(right) The same DMM measuring ohms. For ohm measurements, the meter uses an internal power source to generate the resistance measurements and any outside voltage will cause errors. Therefore, it is necessary to disconnect all power to the test device. The test leads are then placed across the device; the same as when performing a normal voltage measurement.

Fig. 2-15
Making a series measurement with a DMM

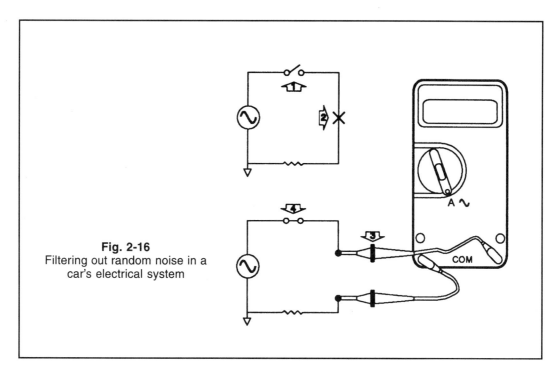

Fig. 2-16
Filtering out random noise in a
car's electrical system

Measuring Current

To measure current, open the circuit so that the meter can be inserted in series (Fig. 2-16). By making the DMM part of the circuit, you can measure the electron flow through the DMM, read in amps. The current shunt has very low resistance and won't cause much voltage drop, often referred to as burden voltage.

Random Noise

Some DC voltages have unwanted noise riding on them. Most modern DMMs have a filter to get rid of this noise. We will later review electrical noise in your car's electrical system, and discover how to solve this problem.

Unraveling Advertised Specifications

Specifications can be confusing and often misleading, instead of being helpful as they should be, but such is the game of advertising! Specifications describe the capabilities of a DMM in terms of measurement capability, resolution, sensitivity, and accuracy. Capacity refers to maximum voltage, current, and resistance which the DMM can measure. Resolution is the degree to which small changes can be detected. Sensitivity is the smallest level of a signal which the DMM can accurately read. Accuracy is the amount of uncertainty stated as a percentage of reading.

Resolution

DMMs are referred to by the number of digits they can display and the total number of counts they can display. They are also referred to by the highest number of counts they can display. As an example, a 2,000-count DMM can display the number 1,999 and a 3½-digit 20,000-count DMM can display 199.7, but if you used a 4½-digit 20,000-count DMM you could display 199. 75.

The 1/2 designation refers to a most significant digit that will display less than all 10 integers, usually a blank or a one. Some DMMs can display more than a blank or a one in the most significant digit (MSD) place. The first digit might be a 1, 2 or 3. DMMs like this measure up to 3,200 counts; and the 3,200-count DMM can give you the same resolution that a typical 4½-digit DMM can give you between readings of 2,000 and 3,200 counts. If this reference to counts has eluded you, refer back to Fig. 2-12 and review how an A/D converter works.

Sensitivity

Sensitivity and resolution are closely related. Sensitivity is the smallest level of a signal that the meter can measure. A DMM that senses changes as small as 100 uV is more sensitive than a DMM which can sense only 100 mV changes. A 4½-digit DMM is not necessarily more sensitive than an ordinary 3½-digit DMM, but most are, and they have 10-times more resolution.

Accuracy

Accuracy is the largest error that will be allowed under specific operating conditions. It is determined by comparing the instrument to national standards, and is usually specified within a range of temperature, humidity, and time. Most of the cost of a low cost DMM goes into meeting a decent accuracy specification. Long calibration cycles are the hallmark of a good quality DMM. That is, the manufacturer guarantees the DMM to be within accuracy tolerances for say 24 or 18 months, not just 90 days!

One factor which can be confusing when you read specifications is that accuracy can be stated in two different ways. These are as a percentage of full-scale readings and secondly as a percentage of the reading made. More specifically, if the accuracy is specified as 1 percent of range, and the range is 200V, then the inaccuracy can be 2V, regardless of the voltage reading. We could be measuring 5V +/- 2V, or 200V +/- 2V. At the bottom of the range, the inaccuracies are substantial, but at the top of the range, they may be acceptable to you. Beware of this potential pitfall! Know the difference between percentage of range and percentage of reading.

Interpreting the Spec Sheet

On a spec sheet, accuracy is shown in percent of reading as:

+/– (1% + 2d)

What does this mean? The first number gives the accuracy in percent. If the true value of a signal is 100V and accuracy is 1 percent, the reading might be between 99V and 101V. The second number represents the range of digits of accuracy. For example, +/- 2 digits of inaccuracy with a reading of l00V indicates a true value between 100.2 V and 99.8 V. This is a range of two digits either way from 100V. We can combine the two and arrive at 100V +/- (1% +2). A true value of 100V could end up on the display as any number from 99.8 to 101.2.

The last digit on the display will often bounce back and forth; this is called rattle. A DMM with a small number in the digit spot in the accuracy specification in the last paragraph will have less rattle, making it easier to read, and giving more accurate readings. With a large number such as 143.7, a one- or two-digit rattle is not significant, but with a smaller number it can be a very bothersome characteristic which will more than annoy you; it will prevent very accurate repeatable DMM measurement results.

Selecting the DMM for Yourself

Now that you know what goes on inside a DMM, and understand the mysteries behind the artful game of specmanship, you may actually want to purchase a DMM. If you have wondered which type is best for your needs, this will help.

The first decision is to decide whether you need a true rms DMM or an averaging DMM. An averaging DMM is good for measuring stable sine waves and undistorted signals which are not rich in harmonics. A true rms DMM is required when you are measuring signals with a wide bandwidth and when measuring square waves with pulse trains from generators in communications systems and also when making measurements within a system with switching power supplies that are becoming ever more common. Averaging DMMs don't respond to the dynamics of these rapid changes. Now comes the icing on the cake or features which you may consider either luxury items or the bare essentials to meet your measurement needs.

AC or DC Coupling

In general, for measuring AC ripple on DC, an AC coupled DMM is required.

Continuity

The audible continuity feature, or beeper, allows for fast troubleshooting multi-wire cables and other applications in which you have to trace signals around in an often cumbersome or tedious path.

Diode Test

Many DMMs have diode test with the better ones providing a constant current source for biasing semiconductor junction, as we have referred to previously.

Relative Reference

The relative reference feature provides relative or offset measurements. The first reading is stored and then subtracted from following measurements and shown as either positive or negative variations from the first measurement.

FUNCTION
Volts AC ~V
Volts DC ⎓ V
Resistance Ω
Milliamps/Amps DC mA A⎓
Frequency   Hz

Table 2-1
Summary of DMM functions

Frequency

Some DMMs, such as the Fluke 8060A, offer frequency measurement capabilities. This DMM will measure signals from 12Hz to 200kHz with resolution of 0.01Hz up to 200Hz. This is more than adequate to measure various rotating objects under the hood.

Peak Hold

This useful function captures and retains the peak voltage of an AC or DC voltage or current surge for as long as you check and record the reading.

Temperature

There are DMMs that measure temperature directly in degrees Celsius (formerly called centigrade) by using a type K thermocouple. The Fluke 8024A and the Beckman HD-140T are examples of this. A type K thermocouple allows measurement of temperatures in the range of -20 to +1,265 degrees Celsius. This is not possible with a solid-state probe. The solid-state temperature sensing probes are good for 100 degrees Celsius maximum. Before studying thermocouples, let's summarize the DMM's features in Table 2-1.

Now that we are familiar with the handy DMM, let's use a typical DMM which has been around for more than 15 years, the Fluke model 77 (Fig. 2-17). The front panel nomenclature (Table 2-2) is probably mostly self-explanatory with the possible exception of K-type thermocouples. This means that the contacts the probes fit into are made of a type of metal suitable for a K-type thermocouple made of a chromel-alumel metal pair fused or otherwise soldered together.

	EQUIPMENT USED IN PERFORMANCE TESTS				
	DMM CALIBRATOR	FUNCTION GENERATOR	POWER AMPLIFIER	RESISTANCE CALIBRATOR	OSCILLOSCOPE
AC Voltage	X		X		
Frequency					
Accuracy		X			
Sensitivity		X			
Trigger Level		X			
DC Voltage	X				
mV DC	X				
Ohms				X	
Diode	X				
mA	X				
A	X				
RPM/Inductive Pickup		X			X

Table 2-2
Front panel nomenclature of a typical modern DMM

A thermocouple is a device consisting of two dissimilar metals joined, usually by soldering or welding. When a temperature difference exists on these metals, a small voltage is created which is proportional to this temperature. It is, therefore, vitally important to know what kind of metal the contacts are since a second set of dissimilar metals will create yet another, even smaller, offset voltage which is an error voltage (Fig. 2-18, a pictorial drawing of the thermocouple probe within the DMM). In our applications section we are going to make so many temperature measurements based on thermocouples we ought to study them a bit more intensely. These are also easy to make if you can solder.

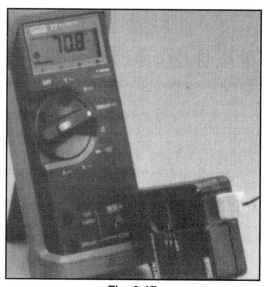

Fig. 2-17
The Fluke DMM has been widely accepted in the aumtomotive repair business.

Thermocouples

There are two types of temperature-sensing probes to be used with a DMM. These are thermocouples and semiconductor junction probes. Selecting the proper probe requires some thought. You need to determine if the substance measured is air, liquid or a surface.

Examining the thermocouple first, this joining of dissimilar metals in the presence of a temperature gradient causes a voltage difference to exist. You can adapt your DMM to accept a thermocouple in two different ways. You can use a special DMM with a reference junction temperature compensation circuit. This is sometimes referred to as a cold junction compensator (Fig. 2-18 again). Or you can purchase a low-cost thermocouple converter assembly. There are several DMMs on the market, such as the Fluke model we are about to examine, which work directly with thermocouples and offer the advantages of temperature measurement capability. Referring again to Fig. 2-18, the isothermal termination connection is used to provide thermal isolation for the thermocouple from the DMM jacks. The K-type thermocouple is made

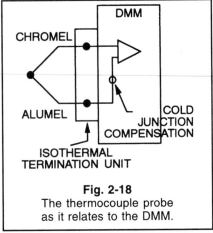

Fig. 2-18
The thermocouple probe as it relates to the DMM.

of the same type materials as the inputs to the DMM. Therefore, as long as the same two types of metals are used throughout the loop, there is no problem, since there are still only two junctions and not additional dissimilar metals come to play into this thermocouple voltage determination scheme.

Actual Applications

Let's start with two simple examples, doorbells and flashlights. Older doorbells used either two or four dry-cell batteries connected in series for 3 or 6V, respectively. Modern units use a step-down transformer for either 10 or 16V AC for doorbells and door chimes, respectively. Figure 2-19 is a typical circuit.

First, check for switch continuity by removing the doorbell switch from its wires and pressing the normally open switch, which should short. The multimeter will

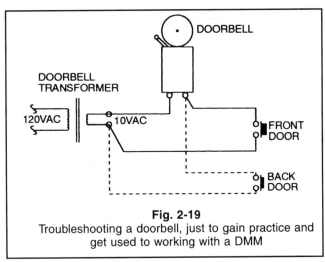

Fig. 2-19
Troubleshooting a doorbell, just to gain practice and get used to working with a DMM

beep. If you have a lighted doorbell the beep will occur without pressing the switch since it has a limiting resistor in series with the switch. In either case, if no beep occurs, disassemble the switch and inspect, then clean or replace it. If the switch is okay, test the AC by placing it across the button leads. A reading in the 10 to 30V AC range should occur; if not, check the transformer or the batteries on the DC range if it is an older doorbell. Refer to Table 2-3 for additional troubleshooting tips.

Component	Possible Causes of Trouble
Doorbell Button	Loose wire terminals or corrosion preventing button from making contact.
Doorbell	Open coil, bad internal commutator switch or loose connection.
Doorbell Circuit Wires	Broken, shorted wires or loose connections
Dry Cell Batteries	Dead batteries or loose connections
Doorbell Transformer	Open or shorted primary or secondary winding, loose connections, no 120V power to primary.

WARNING
DO NOT ATTEMPT TO TROUBLESHOOT OR REPAIR THE 120 VOLT AC LINE PORTION OF THE DOORBELL CIRCUIT UNLESS YOU ARE QUALIFIED TO DO SO.. ANY REPAIRS OR MODIFICATIONS MUST MEET LOCAL ELECTRICAL CODE REQUIREMENTS. CALL A LICENSED ELECTRICIAN IF YOU NEED ASSISTANCE.

Table 2-3
DMM troubleshooting trips

Table 2-4
Battery parameter, shapes, and sizes

Battery Type	Nominal Voltage	Number of Cells	Typical Open Circuit Voltage When New
Size C or D	1.5	1	1.55V
Lantern	6	4	6.40V
Transistor radio	9	6	9.75V

Typical common battery voltage.

Measured Value	Conclusion
1.55V	New
1.49V	Good
1.41V	Poor
1.24V	Dead

Table 2-5
Status or condition of a D cell vs. the voltage measured on it

Typical common battery voltages afte degrees of use (discharge)

Checking flashlight batteries and bulbs is easily done with a DMM. Electrically, a flashlight has only three seriesed parts, a switch, a battery, and a bulb. Carbon-zinc batteries vary in physical sizes, shapes, and voltages. Table 2-4 shows some common batteries. The voltage that a battery produces varies according to the battery's rated current drain. A battery has reached its useful life span, according to manufacturers, when the voltage drops to 80 percent of its new-rated value at full load after a 30-second duration. As an example, if you were testing a D cell, it is rated at 0.15 Amps, so it would drive a 10 ohm load for 30 seconds without falling below 1.2 volts. Table, 2-5 shows typical conditions of a D cell.

Starting Simply

Your car's incandescent light bulbs are either good or bad; there is no intermediate stage. You check this with a continuity check, as you also do with switches. If the switch is not shorting making contact, it causes your DMM to fail to make the appropriate beep. When this happens, dismantle and clean the switch using steel wool and denatured alcohol, or spray it with TV tuner contact cleaner. It is crucial for the impedance to match that of your your car's speakers; mismatches cause distortion, overheating, and potential damage. Most speakers are either 4, 8, or 16 ohms with 8 ohms the most common value. The output amplifier and speaker inputs are usually labeled; however, if they are not, you can measure the impedance as follows:

■ With the DMM in the resistance mode, disconnect any wires attached to the speaker terminals and hold the leads against the terminals and record the reading. The reading is DC resistance as is typically 80 percent of the rated impedance.

Phasing

This means that both speakers work together pushing air in the same direction at the same time. If improperly phased, these pressures cancel one another out and the realism of stereo is reduced. They are normally marked as either "+" and "−" or red and black. In either case connect like colors and/or algebraic signs together. If one has colors and the other has a sign, black equates to a "−" sign. If these markings are absent, or if you want to double-check them, follow the following procedure:

1. Set the meter to the CONTINUITY mode and disconnect the speaker cables from both the amplifier and speakers.
2. Use the amplifier end of the cable as a starting reference. For a twisted pair cable, choose one wire and label it "+". For a pin jack cable, use the center pin. Bring the speaker end close enough so that you can reach both ends with the test leads. Hold one test lead to the end labeled "+" while touching the other test lead to each wire at the speaker end. Label it in the same way as the amplifier end and repeat the test for the other speaker cable. Now connect the speakers as earlier mentioned with the speaker's "+" terminal to the amplifier's "+" terminal, etc.

Digital multimeters (DMMs) are transformed into digital thermometers when you use a temperature-sensing accessory. In this application, they can help minimize energy consumption and maximize comfort by detecting cold-air leaks, measuring relative humidities, and ensuring correct temperature adjustments are made on appliances.

Measuring Temperature

A modern DMM accepts two forms of heat sensors, either a thermocouple or a semiconductor heat sensing probe. The thermocouple-type probe measures higher temperatures, -150°C to 1,000°C. Both convert heat to a voltage, so the DMM is actually displaying voltage when it senses heat. With these devices, 10C or 10F equals exactly 10mV so no "scaling" or multiplying by a non-decimal number is required. Some sensors have a selector switch which internally scales the voltage gain of an amplifier to display OC or OF.

Measuring Temperature Gradients

This starts by measuring the temperature in, for example, the center of the radiator. Then examine potentials for leaks around casings, pressure outlets, and then measure the radiator's surface itself. Record these, making a profile of readings across your cooling system. Ideally, your cooling system should have the same temperature in all parts of each of its components. However, in the laws of physics relating to temperature and heat transfer, there are thermal lags. In other words, the heat entering one component in this cooling system chain can't vent its heat to the outside as well as other components. If the radiator were maybe 50 times larger, this differential would be greatly reduced; however, this is naturally impractical.

Relative Humidity

This is important in maintaining a comfortable house and car. The best range is between 35 percent and 65 percent. Lower than this, your throat will feel dry, static electricity is present, and your skin will crack. Above this range, condensation forms on your car's windows. There also is a greater tendency for your car to rust.

To control moisture, look at its sources. Moisture control requires good venting and insulation with a moisture barrier. This should be placed between the heated or cooled surfaces in your car. Also, a sheet of plastic over your car at night will minimize the amount of moisture entering your car. That is why we have garages. But how do you measure moisture?

Following the Professionals

Heating and air conditioning professionals use a device called a "sling psychrometer," which consists of two glass thermometers mounted on a swivel handle. This device allows the thermometers to spin around, providing air motion. One thermometer has a wick over the bulb, which is saturated with water to indicate wet bulb or dew point temperature, which we duplicate here.

The Measurement Procedure

This begins by taking a shoelace from an old tennis shoe and cutting it to a 2½-inch length. This nonsolid type strand has a hollow cross woven center. Take this and wet it, and slip it over the thermocouple and hold it level. Take a newspaper and fan it, but do not blow on it because your breath is warmer and more moist than the air you are measuring. Next, repeat this procedure without the "sock" shoelace and with a dry thermocouple tip. Be sure to fan it again. Take the dry and wet readings and refer to Fig. 2-20, the psychrometric chart, and plot the wet reading along the curved axis and the dry reading along the horizontal axis. Note the example in this figure for a wet bulb temperature of 17°C and a dry bulb temperature of 24°C. Where they intersect is the relative humidity.

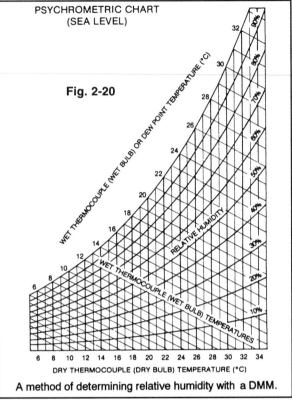

Fig. 2-20

A method of determining relative humidity with a DMM.

Measuring temperatures with an ordinary mercury-in-glass thermometer is risky since if it breaks, it contaminates the portion of your car it rests on. A better solution is the far more rugged thermocouple sensor.

Thermocouple Sensor Applications

Constructing a Probe

Begin by using a steel-tube casing from a junk automobile's brake lines. Next, drill a small hole in a 2" x 2" block of wood that will serve as a handle. Drive the handle onto the end of the tubing, then slightly flatten the end of the tubing opposite the handle (Fig. 2-21). Insert the thermocouple tip into the hole in the handle and push it through until the measuring tip extends 1½ inches beyond the flattened tubing tip. Insert the flattened end of this "probe" tip

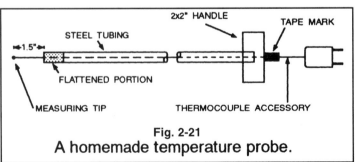

Fig. 2-21
A homemade temperature probe.

into the center of a bale of hay, then withdraw it about 2 inches. Read the temperature in its center, realizing that the temperature shouldn't be stored if it is 1°F greater than the temperature at a depth of 3 inches into the bale.

Constructing an even longer probe is needed when you are measuring hard-to-access temperatures. Using thermocouple wire, such as the Fluke model Y8111 insulated thermocouple wire, and the matching connector pair (Fluke model Y8115) extends this to 100 feet, far beyond what you would ever need. In these applications, you may want to "bury" several thermocouples so that you can make "layered" temperature measurements, realizing that temperature and moisture are proportional.

Using tubing such as with the hay bale application, insert the temperature probe into the soil and leave the tube inserted into the ground and mark and cover it from rain which will distort the temperature measurement.

Measuring Outdoor Temperatures

Many cars have an indoor display that indicates outdoor temperature. It is easy today, especially with modern DMMs, that record the lowest and highest readings in memory, to very accurately check this. When making outside measurements, try to position the thermocouple so it does not sense the heat coming off your engine or exhaust system.

Using a DMM to Check MOSFETs

If you want to really become adventuresome, attempt to fix car stereos and radios. Their most common faults are burned-out output transistors, which you can check. A DMM's diode-test function enables you to test power MOSFETs. The potential across the DMM's probes in this mode is enough to exceed the gate-to-source threshold of most enhancement-type MOSFETs. And, due to a MOSFET's gate-source leakage being so very low, its open-circuit gate will hold a charge long enough for you to make certain measurements.

To test an N-channel FET (the majority of these transistors are N-channel), attach the DMM's negative probe to the FET's source terminal and the positive probe to the FET's gate (Fig. 2-22a). The DMM will indicate an open circuit for a normal device. Next, remove the positive probe, while being very careful not to remove the negative probe from the FET's source. If you fail to maintain contact with the source you will have the probe acting exactly like an antenna. Once you have removed the positive probe from the FET's gate, connect it to the drain (Fig. 2-22b). The DMM will read 0 Volts for a normal device.

For the final test, short the gate-to-source capacitance (Fig. 2-22c). The DMM should show an open circuit for a normal device. For a P-channel FET repeat this procedure only with the leads reversed.

Let's conclude by testing your thermostat and radiator surface temperature. If you can manage to slip a thin wire down the radiator with the pressure cap on, you can also measure the radiator's far hotter interior temperature. To test the thermostat, you first need to know what its trip temperature is (Fig. 2-6. and note the 192 in this illustration). This is its temperature (in

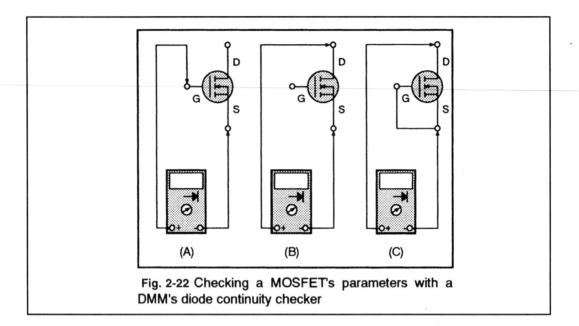

(A)   (B)   (C)

Fig. 2-22 Checking a MOSFET's parameters with a DMM's diode continuity checker

degrees F) at which it will close. You may want to immerse it in very hot water and boil it to see when it closes. You might want to also measure the temperature at which it closes to check it against the manufacturer's data sheet or claims since these can vary by +/– 5% easily. Under these circumstances, measure the resistance across this open-to-close gap. It should be less than 1 Ohm, often indicated as OL or low resistance on your DMM. Since we have already covered temperature measurements with a thermocouple, you can now apply this to your radiator. As a note: The Fluke model 77 and 88 DMMs are both made especially for troubleshooting your car. They are also very nearly identical. The model 88, though, has a few extra features, but the less expensive model 77 will do a nice job on anything that you require of it.

Chapter 2 Quiz

True or False

1. Cooling system hazards are more common than flats.
2. More than 50 percent of cars undergoing an engine change also have their cooling systems reworked.
3. Your "cooling system" is a slight misnomer.
4. There is very little metal-to-metal rubbing within an internal combustion engine
5. When your radiator vents heat to the outside world, this is a form of heat exchange.
6. There are three methods of heat transfer.
7. In a typical car, your car's coolant circulates from 2,000 to 7,000 gallons/hr.
8. Engine overheating breaks down your antifreeze, causing it to become more acidic.
9. Most coolant mixtures contain about 90 percent antifreeze.
10. When antifreeze breaks down, it eats more metal and causes more leaky seals.

Matching

Enter the correct letter after each sentence or statement

11. You need to do this when your radiator springs a leak.
12. These metals usually comprise a radiator.
13. One of the three methods of heat transfer.
14. This generates enough thermal energy to heat a 5-room house to 70 degrees F when it is just 0 degrees F outside.
15. This is your cooling system's worst culprit.
16. This is the phenomenon of a metal expanding, as it becomes hotter.
17. Great pressure builds up within your radiator, requiring this.
18. One of two kinds of materials radiator hose are made from.
19. This is part of the outlet hose.
20. One of two kinds of radiator clamps.
21. One of two kinds of fan clutches for a variable pitch fan.
22. Your water pump pulley usually drives this.
23. This is an added task of the drive belt, along with turning your fan blades.
24. This device is for pressure relief so your radiator doesn't burst when its coolant freezes.
25. This device resides between the front of your engine and your radiator's top or "inlet" hose.
26. This parameter varies from 160 degrees F to 220 degrees F.
27. This is the connector between the pressure cap and the top of your radiator.
28. This is an alternate avenue by which your coolant pours into your radiator.
29. This device attaches to the side of a filler neck.
30. One of three kinds of overflow tubing.

A. Brass or Copper

B. The radiator filler neck

C. The freeze plug

D. Have it welded

E. The surge tank

F. Copper, steel or a flexible compound

G. Conduction, convection or radiation

H. A positive temperature coefficient

I. It turns your alternator

J. A spiral shaped wire

K. The thermal trip point of a thermostat

L. Molded rubber or an elastomer

M. Overflow tubing

N. A thermostat

O. Fan blades

P. An obstruction in the circulation path

Q. A pressure cap

R. A wire or screw type

S. Centrifugal or thermostatically controlled

T. Either your engine, exhaust or radiator

Fill in the Blanks

31. The beeper function on your DMM indicates a _____ or a closed circuit.

32. The frequency counter on your DMM is ideal for measuring _____ objects or shafts.

33. In order to prevent you from unsoldering components on a PC board, many modern DMMs have a very low _____ voltage.

34. The Fluke model 88 has a test voltage of _____.

35. The current shunt in your DMM has very little resistance and therefore won't cause much of a _____ drop.

36. Many modern DMMs have a filter to get rid of unwanted electrical _____.

Multiple Choice

37. Newer DMMs have:

A. Frequency counters built in

B. Peak hold

C. Conductance

D. All of the above

38. The analog DMM

A. Require shunt resistors

B. Never "ping" their needles

C. Use current generated by an internal battery to measure resistance

D. A and C

39. The discharge rate of circuitry attached to your DMM is always constant due to an internal capacitor interacting with a

A. Transistor

B. Op Amp

C. Resistor

D. Another capacitor

40. This parameter describes the smallest detectable level you can measure with your DMM
A. Frequency
B. Conductance
C. Sensitivity
D. Resolution

41. You some times can't use a semiconductor temperature sensor in automotive trouble-shooting because
A. It can't withstand temperatures above the boiling point of water
B. They are too large
C. They are too slow to react
D. They are too expensive

42. A thermocouple
A. Consists of two pieces of dissimilar metal (usually) in the form of wires joined at one point
B. Is never accurate
C. Can measure very high temperatures
D. A and C

43. When working on your car's electronic entertainment devices, it is necessary to properly phase the speakers because
A. It detracts from the stereo effect
B. It enhances the stereo effect
C. It enhance channel separation
D. Both A and C

44. Ideally (both not realize in practical terms), your cooling system should have
A. Total uniformity in its heat distribution (the same temperature everywhere)
B. Temperatures always less than your radiator
C. No antifreeze used in summer, only water
D. None of the above

45. You can improvise a device for measuring relative humidity by using
A. A tennis shoe
B. A tennis shoe's lace
C. A 2" x 2" block of wood
D. Both B and C

46. You can build your own temperature probe by
A. Using a square small block of wood
B. A and C
C. Auto brake lines
D. None

47. You can check your thermostat by realizing
A. It shorts electrically when it physically closes
B. It closes at or around its trip temperature's threshold
C. You'll use either your continuity beeper or ohm scale to measure it
D. All of the above

48. You measure relative humidity to
A. Determine potential spots of further rust as well as the likelihood of moisture seepage that eventually causes rust
B. Determine its method of drainage when hot dry weather comes
C. Know where to place plastic sheets
D. None of the above

3

More Involved and Advanced DMM Measurements on Your Car's Electrical System

In Chapter 2 you began to get a feel for what a DMM could accomplish, how it functions, and see how it has some limitation. Chapter 3 logically extends this general DMM knowledge we gained. It delves into the realm of actually troubleshooting some advanced features on your own car with a Fluke Model 88 DMM. This is a DMM solely designed with one purpose in mind— troubleshooting your automobile, primarily the electrical systems.

Figure 3-1 shows the Fluke Model 88 automotive DMM. You will see a number of symbols on it that Table 3-1 explains. The Model 88 has four input terminals (see Fig. 3-2). These are protected against overloads to the extent shown in the illustration. If you place a test lead into either terminal **A** or **mA**, and the selector switch is not set to **mA**, **A**, DC or AC, the beeper emits an input alert that sounds like a clicking noise. You can disable this input alert by holding down the **ALERT** button while turning the rotary switch from OFF to any other position. You should never attempt to make a voltage measurement if a test lead is in the Amp (**A**) or milliamp (**mA**) position!

The eight-position rotary selector switch (see Fig. 3-3) selects seven functions, plus the OFF position. Table 3-2 shows the unit symbols you can measure with this DMM. The DMM's entire display lights as part of a self-test initiation routine. After this, the DMM is ready to use. If you press and continuously hold down any pushbutton while turning the meter from off to on, the display remains lit until you release the pushbutton. Table 3-3 lists the functions of each of this DMM's eight pushbuttons. The pushbuttons select various operations (see Fig. 3-4). When you push a pushbutton, a display symbol lights, and the beeper sounds (unless you have turned off all beeper functions). Turning the rotary switch to another switch setting resets

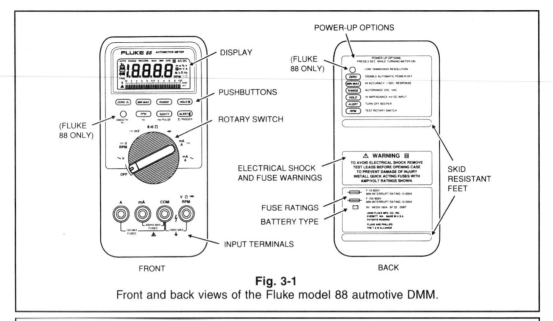

Fig. 3-1
Front and back views of the Fluke model 88 autmotive DMM.

Table 3-1
Symbols associated with, and safety practices that should be follwed with a DMM.

⚡	DANGEROUS VOLTAGE	⏚	GROUND
∼	AC-ALTERNATING CURRENT	⚠	SEE EXPLANATION IN MANUAL
⎓	DC-DIRECT CURRENT	▣	DOUBLE INSULATION (Procection Class)
∿	EITHER DC OR AC	⌑	FUSE

USING YOUR METER SAFELY

Follow safe servicing practices as described in your automobile service manual. To ensure that you use the meter safely, follow the safety guidelines listed below:

- Avoid working alone.

- Disconnect the power and discharge all high-voltage capacitors before testing in the ⏜Ω and ⊣⊢ functions.

- Inspect the test leads for damaged insulation or exposed metal. Check test lead continuity. Damaged leads should be replaced.

- Do not use the Meter if it looks damaged.

- Select the proper function and range for your measurement.

- Use caution when working above 60V dc or 25V ac RMS. Such voltages pose a shock hazard.

- When using the probes, keep your fingers away from probe contacts. Keep your fingers behind the finger guards on the probes.

- Disconnect the live test lead before disconnecting the common test lead.

- When measuring current, turn the power off before connecting the Meter in the circuit.

- Check Meter fuses before measuring current transformer secondary or motor winding current. (See "Testing the Fuses" in the "MAINTENANCE" Section.) An open fuse may allow high voltage build-up, which is potentially hazardous.

- Use clamp-on probes (current clamps) when measuring circuits exceeding 10A.

Measurement Unit Symbols

SYMBOL	MEANING
AC	Alternating current or voltage
DC	Direct current or voltage
V	Volts
mV	Millivolts (1/1000 volts)
A	Ampere (amps). Current
mA	Milliampere (1/1000 amps)
%	Percent (for duty cycle readings only)
Ω	Ohms. Resistance
kΩ	Kilohm (1000 ohms). Resistance
MΩ	Megohm (1,000,000 ohms). Resistance
Hz	Hertz (1 cycle/sec). Frequency
kHz	Kilohertz (1000 cycles/sec). Frequency
RPM 1	Revolutions/minute. Counting one cycle per spark.
RPM 2	Revolutions/minute. Counting 2 cycles per spark
ms	Milliseconds (1/1000 sec) for Pulse Width measurements.

Table 3-2
Unit symbols and their meanings.

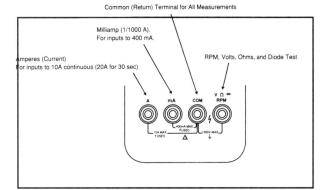

Fig. 3-2
The model 88's input terminals.

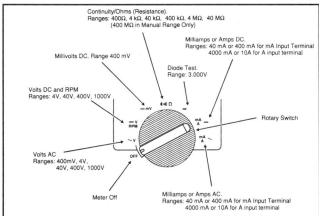

Fig. 3-3
The model 88's selector rotary front panel switch.

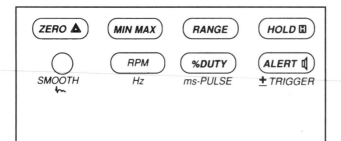

Fig. 3-4
The eight pushbuttons and their
functions, on the model 88.

Pushbutton Operations

PUSHBUTTON	FUNCTION	DISPLAY SYMBOLS
ZERO △	ZERO (Relative Reading) Function Displays difference between the measured value and the stored value.	△
MIN MAX	Minimum (MIN), Maximum (MAX), Average (AVG) Recording. Records minimum, maximum, & calculates true average.	RECORD, MAX, MIN, AVG
RANGE	Manual Range or Autorange In Manual Range user selects fixed range. Meter stays in that range until user changes it, selects autorange, or turns Meter off. In Autorange Meter selects range automatically.	RANGE, AUTO
HOLD Ⓗ	Touch Hold Touch Hold holds last stable reading on display. A new stable reading, causes beeper to sound and display to update. If Meter is in MIN MAX Recording, RPM, Duty Cycle, Pulse Width, or Hz, Touch Hold interrupts the function. Display is frozen, but recorded readings are not erased. Pressing MIN MAX, ALERT, or SMOOTH when meter is in Touch Hold causes Meter to exit Touch Hold and enter MIN MAX Recording, change alert, or SMOOTH.	Ⓗ
RPM Hz	RPM 2, RPM 1, or Frequency RPM 2. 4-cycle engines, not DIS (counts 2 cycles/spark). RPM 1. 2-cycle engines and DIS (counts 1 cycle/spark). Hz counts frequency between 0.5 Hz and 200 kHz.	RPM①, RPM② Hz
% DUTY ms-PULSE	Duty Cycle or Pulse Width Duty Cycle between 0.0-99.9% displayed. Pulse Width between 0.002-1999.9 ms displayed.	%, ms
ALERT ±TRIGGER	Change Alert, Continuity Beeper, or ± Trigger In voltage or current function selects Change Alert. In Ω function selects Continuity Test. In Duty Cycle or Pulse Width selects trigger slope.	◁•))) , TRIG, +, -
SMOOTH	SMOOTHING Function and Back-light Display (Fluke 88 Only) SMOOTH displays average of last eight readings. Press YELLOW button to turn on or off display back-light. (Backlight turns off automatically after 68 sec.)	∿

Table 3-3
Functions and meanings of the model 88's pushbuttons.

all pushbuttons to their default state. The Model 88 automotive DMM has both digital and analog displays (see Fig. 3-5). If the inputs are stable, the digital display offers the best choice of viewing and most accurate reading. However, if you have a continuously varying quantity, you can read the analog pointer.

If you apply an input that is too large, **OL** (overload) shows on the display. Also, if you are attempting to make a duty cycle measurement and the signal stays either high or low, refusing to switch states, an **OL** also appears on the display. The display symbols indicate what the DMM

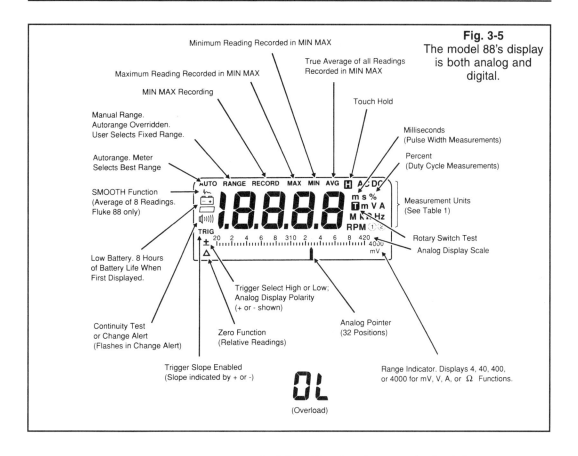

Fig. 3-5
The model 88's display is both analog and digital.

Minimum Reading Recorded in MIN MAX

Maximum Reading Recorded in MIN MAX

True Average of all Readings Recorded in MIN MAX

MIN MAX Recording

Touch Hold

Manual Range. Autorange Overridden. User Selects Fixed Range.

Milliseconds (Pulse Width Measurements)

Autorange. Meter Selects Best Range

Percent (Duty Cycle Measurements)

SMOOTH Function (Average of 8 Readings. Fluke 88 only)

Measurement Units (See Table 1)

Low Battery. 8 Hours of Battery Life When First Displayed.

Rotary Switch Test

Analog Display Scale

Trigger Select High or Low; Analog Display Polarity (+ or - shown)

Continuity Test or Change Alert (Flashes in Change Alert)

Zero Function (Relative Readings)

Analog Pointer (32 Positions)

Trigger Slope Enabled (Slope indicated by + or -)

Range Indicator. Displays 4, 40, 400, or 4000 for mV, V, A, or Ω Functions.

(Overload)

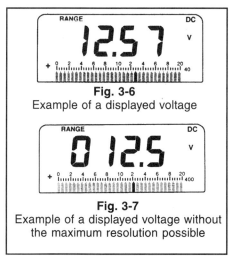

Fig. 3-6
Example of a displayed voltage

Fig. 3-7
Example of a displayed voltage without the maximum resolution possible

is doing. Keeping the meter in the right measurement range is important since selecting too low of a range will cause the Model 88 to display **OL**. Selecting too high a range will not allow you to take full advantage of the Model 88's accuracy. On this DMM, you select the ranges in multiples of four, which are 4.000V, 40V, 400V, then 1000V maximum. As an example, if you were to select the 40V range, the DMM reads 12.57 V on the display (see Fig. 3-6). If you press RANGE once, you place the DMM in the 400.0V range. Since the 400.0V range is too high for the 12.57V input, the Model 88 DMM drops the last digit and displays the measurement in a less accurate fashion (see Fig. 3-7). If you were to press RANGE twice more, you'd wrap to the 4.000V range and the DMM would display overload: **OL**.

The Model 88 has Auto-range and Manual range operations. When you select the DC volts function, the DMM defaults to the 40V DC range. When you select the AC volts function, the DMM defaults to the 4V AC range. When you select the amp or ohms range, the DMM automatically defaults to Autorange. In Autorange (AUTO), the DMM selects the best range. This allows you to switch from one test point to another without having to worry about resetting your range. In the Manual range (RANGE), you select the range yourself. This naturally allows you to override Autorange. The DMM stays in this range until you direct it to Autorange. If you manually change the measurement range after you enter TouchHold™, MIN MAX Recording, and/or ZERO, the DMM exits these functions. When in the Manual range, pressing RANGE steps up a range. In the highest range, the DMM wraps to the lowest range. To switch from Manual to Autorange, you have to press and hold down RANGE for two seconds. To switch back, you have to press RANGE. The DMM enters Manual range in the range in which it already is at that moment.

Testing Your Car Battery

Figure 3-8 shows the setup to measure your car battery's no-load voltage. As you may recall, we did this as an example of gaining a better understanding of the concept of voltage in Chapter 1. The chart within Figure 3-8 shows the voltage that corresponds to various states of charge of the battery. This procedure is straightforward:

1. Insert the Model 88 DMM's red test lead in the **V** input terminal, and then insert the black lead in the **COM** (common) terminal.
2. Set the rotary switch to **V DC**.
3. Turn on your headlights for one minute to bleed off surface charge.
4. Turn lights off and touch the probes to the circuit as shown in the figure. This places the DMM in parallel with the circuit. Voltage must always be measured with the DMM in parallel with the circuit.
5. Read the display. A fully charged battery typically shows about 12.6V.

The no-load voltage indicates the state of charge, not the condition of the battery. A weak battery may indicate a full terminal voltage when it is not supplying current to some accessory.

Testing Diodes on Alternators

Since a good diode permits current flow in one direction only, you'll have to remove the diode from the circuit and turn power off. We'll perform this test in a conventional manner and then with the beeper function on the DMM. The conventional method is as follows (see Fig. 3-9):

1. Disconnect the battery cable from the alternator output terminal.
2. Insert the red test lead in the V **W** input terminal. Insert the black test lead in the **COM** terminal.
3. Set the rotary selector switch to the diode symbol.
4. Touch the red test lead to chassis ground and the black test lead to the alternator output terminal. Read the display.

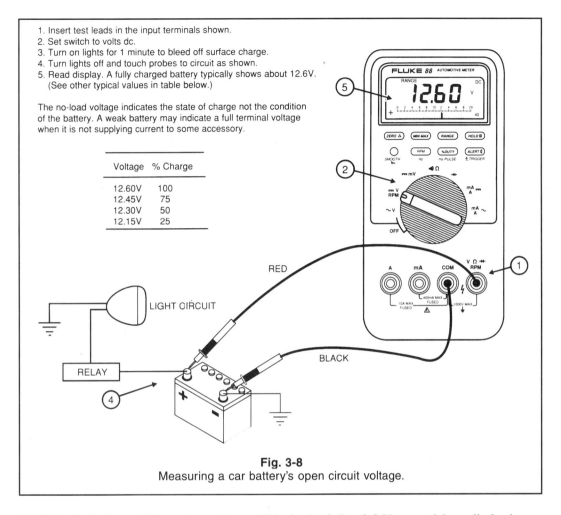

1. Insert test leads in the input terminals shown.
2. Set switch to volts dc.
3. Turn on lights for 1 minute to bleed off surface charge.
4. Turn lights off and touch probes to circuit as shown.
5. Read display. A fully charged battery typically shows about 12.6V.
   (See other typical values in table below.)

The no-load voltage indicates the state of charge not the condition of the battery. A weak battery may indicate a full terminal voltage when it is not supplying current to some accessory.

Voltage	% Charge
12.60V	100
12.45V	75
12.30V	50
12.15V	25

Fig. 3-8
Measuring a car battery's open circuit voltage.

5. If the diode is good, the voltage drop will be in the 0.5 to 0.8 V range. Many diodes in use drop 0.3V (germanium diodes to be specific). Typically, two alternator diodes in series drop 0.8 V. NOTE: There are two types of semiconductors and diodes. These are silicon and germanium. Their main differences are that, almost regardless of the current flowing through them, a silicon diode drops 0.7 volts and a germanium diode drops 0.3 volts (see Chapter 2).

6. Reverse the test leads and measure the voltage across the diode again. If the diode is good, the DMM displays **OL**.

If the diode is shorted, it displays "O" in both directions. If the diode displays **OL** in both directions, it is open. To perform this diode test using the DMM's beeper function, you would proceed as follows:

A. Perform steps 1 and 2 again, then press **HOLD** to enter Touch Hold™.

B. Touch the red test lead to the positive side and the black lead to the negative side.

C. Reverse the probes.

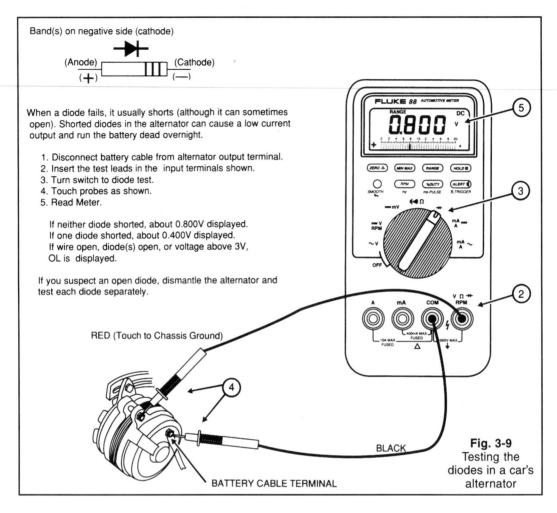

Band(s) on negative side (cathode)

(Anode)   (Cathode)
(+)   (—)

When a diode fails, it usually shorts (although it can sometimes open). Shorted diodes in the alternator can cause a low current output and run the battery dead overnight.

1. Disconnect battery cable from alternator output terminal.
2. Insert the test leads in the input terminals shown.
3. Turn switch to diode test.
4. Touch probes as shown.
5. Read Meter.

If neither diode shorted, about 0.800V displayed.
If one diode shorted, about 0.400V displayed.
If wire open, diode(s) open, or voltage above 3V, OL is displayed.

If you suspect an open diode, dismantle the alternator and test each diode separately.

RED (Touch to Chassis Ground)

BLACK

BATTERY CABLE TERMINAL

Fig. 3-9
Testing the diodes in a car's alternator

If the diode is good, the DMM beeps when the test leads are in one position and remain silent when they are reversed. If the diode is shorted (or the resistance is less than 4kW), the DMM beeps in both directions.

If the diode is open, the DMM does not beep in either direction. A beep is only activated when a low resistance condition indicates a short or a closure of a continuity path. Be aware that these procedures are not valid for the avalanche diodes found on late-model GM cars. An avalanche diode is one in which the reverse current run causes it to break down at a predetermined exact voltage.

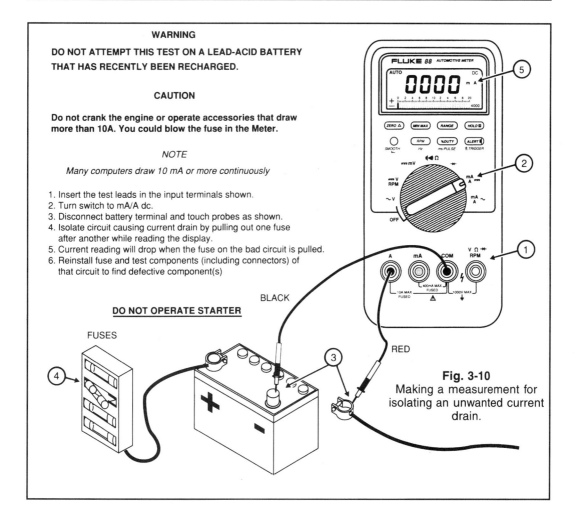

WARNING

DO NOT ATTEMPT THIS TEST ON A LEAD-ACID BATTERY THAT HAS RECENTLY BEEN RECHARGED.

CAUTION

Do not crank the engine or operate accessories that draw more than 10A. You could blow the fuse in the Meter.

NOTE

Many computers draw 10 mA or more continuously

1. Insert the test leads in the input terminals shown.
2. Turn switch to mA/A dc.
3. Disconnect battery terminal and touch probes as shown.
4. Isolate circuit causing current drain by pulling out one fuse after another while reading the display.
5. Current reading will drop when the fuse on the bad circuit is pulled.
6. Reinstall fuse and test components (including connectors) of that circuit to find defective component(s)

DO NOT OPERATE STARTER

FUSES

BLACK

RED

Fig. 3-10
Making a measurement for isolating an unwanted current drain.

Isolating Unwanted Current Drains

Before you attempt to fault isolate this drain, there are two important considerations or actions you ***absolutely must*** avoid. The first is not to try this on a very recently recharged lead acid battery. The second warning is to make certain that the power is off before unsoldering components, cutting conductors or wires, or just generally disconnecting a circuit. This is especially true of ones that carry more than 10 Amps. You can tell this by their fuse rating, which would also be in excess of 10 Amps. Cranking the engine or operating electrical accessories that draw more than 10 Amps can blow a fuse and damage your DMM, and also damage your car's wiring as well.

Lastly, remember that to keep track of such things as miles per gallon, average speed per hour, and related issues, on-board computers must have at least their dynamic RAM memories energized continuously. This type of memory is volatile, which means it must have constant power applied to it or else its contents will be inadvertently erased. Realizing this, do not be surprised to discover a constant 10 **mA** or greater current drain, which is a legitimate operating condition. Don't attempt to disconnect this or your entire electronic dash will have to be reset. If you have ever changed batteries you will know the hassle you face after changing one. That is why some shops hook a good battery in parallel with a faulty battery during the switching process. Figure 3-10 shows the test setup. The procedure is as follows:

1. Insert the red test lead in the **A** or **mA** input terminal and the black test lead in the **COM** input terminal. To avoid blowing a fuse, use the 10 **A** input terminal unless you are absolutely certain the current drawn is less than 400 **mA**.
2. Set the rotary selector switch to **mA/A** (AC or DC). If you set the switch to volts AC or DC, **W** or diode (and the beeper has not been disabled), the beeper emits a rapid clicking sound (input alert).
3. Disconnect the clamp from the negative battery terminal and place the probes as shown in the figure. This places the DMM in series with the circuit being tested, causing all current to flow through the DMM. Current always is measured with the DMM in series with the circuit.
4. Isolate the circuit causing the current drain by pulling out one fuse after another while reading the display.
5. The current reading will drop when the fuse on the bad circuit is pulled.
6. Reinstall the fuse and test components (including connectors) of that circuit to find the defective component.

Testing Circuit Continuity

Figure 3-11 shows how to perform a continuity test with this DMM. This test verifies you have a closed circuit. The DMM can detect opens or shorts with durations as small as 1 ms. This proves very valuable in troubleshooting intermittent problems associated with cables, switches, and relays. We'll use the beeper function on the DMM for this since it was especially designed for this situation. The procedure is as follows:

1. Insert the test leads in the V **W** and **COM** input terminals.
2. Set the DMM's rotary switch to continuity test.
3. Press **ALERT**. The symbol appears on the display.
4. Touch the test leads to the circuit, in this case a typical stoplight switch.
5. Press the brake pedal and listen for the beeper. A continuous tone confirms that you have circuit continuity. Incidentally, a beeper tone means a low resistance, not necessarily zero ohms.

Table 3-4 shows the beeper response in continuity tests and the resistance thresholds of various ranges.

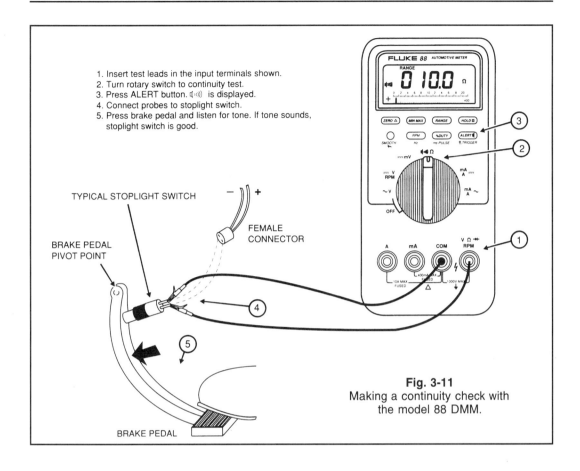

1. Insert test leads in the input terminals shown.
2. Turn rotary switch to continuity test.
3. Press ALERT button. ⊄·))) is displayed.
4. Connect probes to stoplight switch.
5. Press brake pedal and listen for tone. If tone sounds, stoplight switch is good.

TYPICAL STOPLIGHT SWITCH

FEMALE CONNECTOR

BRAKE PEDAL PIVOT POINT

BRAKE PEDAL

Fig. 3-11
Making a continuity check with
the model 88 DMM.

Beeper Responses in Continuity Test

Input Range	Beeper On @*
400.0Ω	less than 40Ω
4.000 kΩ	less than 200Ω
40.00 kΩ	less than 2 kΩ
400.0 kΩ	less than 20 kΩ
4.000 MΩ	less than 200 kΩ
40.00 MΩ	less than 200 kΩ

Table 3-4
The DMM beeper
response's thresholds.

* Below these resistances the beeper will always sound. The beeper can sound at resistances above those indicated.

Resistance Measurements

These tests are not to be confused or co-mingled with continuity tests. Resistance, where it is not supposed to be, hinders the flow of current. Excessive resistance restricts the proper intended amount of current flow. If your headlights don't receive enough current, they will not shine as brightly as the manufacturer intended. Before testing you must turn off the power to the circuit under test. If an external voltage is present across a component, it is impossible to take an accurate measurement of the resistance of that component. The procedure for making resistance measurements on the ignition coil follows (see Fig. 3-12):

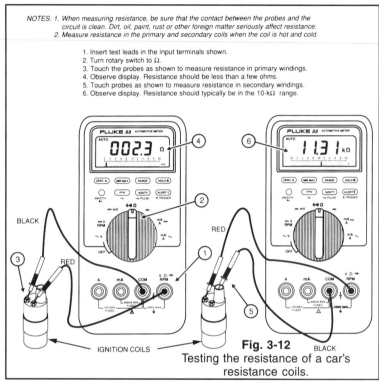

NOTES: 1. When measuring resistance, be sure that the contact between the probes and the circuit is clean. Dirt, oil, paint, rust or other foreign matter seriously affect resistance.
2. Measure resistance in the primary and secondary coils when the coil is hot and cold.

1. Insert test leads in the input terminals shown.
2. Turn rotary switch to Ω.
3. Touch the probes as shown to measure resistance in primary windings.
4. Observe display. Resistance should be less than a few ohms.
5. Touch probes as shown to measure resistance in secondary windings.
6. Observe display. Resistance should typically be in the 10-kΩ range.

BLACK · RED · RED · IGNITION COILS

Fig. 3-12 Testing the resistance of a car's resistance coils. BLACK

1. Insert the test leads into the V **W** and **COM** input terminals.
2. Set the DMM's selector rotary switch to **W**.
3. Touch the probes as shown in the figure to measure resistance in primary windings.
4. Observe the display. Resistance should be less than a few ohms.
5. Touch probes as shown in the figure to measure resistance in secondary windings.
6. Observe the display. Resistance should typically be in the 10 kW range.

The Fluke Model 88 DMM has a Lo Ohms/High Resolution option you can select on power up, which allows you to measure resistance as low as 0.01 **W**. Accurately measuring this low of a resistance is not possible with any ordinary DMM, in fact, 0.2 **W** is good for most DMMs. Later we will explain how to zero out test leads before performing this very low resistance test. Resistance in standard test leads is about 0.1 to 0.2 **W**.

Combating Noise

Rapidly changing display readings (noise) can sometimes be eliminated if you change to a higher scale (range). You can smooth out noise somewhat by using the inductive pickup that comes with this test/troubleshooting instrument. The averaging (AVG) feature of the MIN

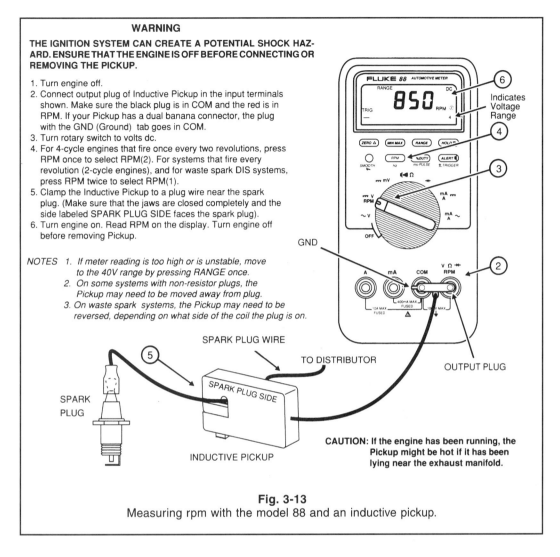

WARNING

THE IGNITION SYSTEM CAN CREATE A POTENTIAL SHOCK HAZ-
ARD. ENSURE THAT THE ENGINE IS OFF BEFORE CONNECTING OR
REMOVING THE PICKUP.

1. Turn engine off.
2. Connect output plug of Inductive Pickup in the input terminals
   shown. Make sure the black plug is in COM and the red is in
   RPM. If your Pickup has a dual banana connector, the plug
   with the GND (Ground) tab goes in COM.
3. Turn rotary switch to volts dc.
4. For 4-cycle engines that fire once every two revolutions, press
   RPM once to select RPM(2). For systems that fire every
   revolution (2-cycle engines), and for waste spark DIS systems,
   press RPM twice to select RPM(1).
5. Clamp the Inductive Pickup to a plug wire near the spark
   plug. (Make sure that the jaws are closed completely and the
   side labeled SPARK PLUG SIDE faces the spark plug).
6. Turn engine on. Read RPM on the display. Turn engine off
   before removing Pickup.

NOTES   1. If meter reading is too high or is unstable, move
            to the 40V range by pressing RANGE once.
        2. On some systems with non-resistor plugs, the
            Pickup may need to be moved away from plug.
        3. On waste spark systems, the Pickup may need to be
            reversed, depending on what side of the coil the plug is on.

Indicates
Voltage
Range

GND

SPARK PLUG WIRE

TO DISTRIBUTOR

OUTPUT PLUG

SPARK PLUG SIDE

SPARK
PLUG

CAUTION: If the engine has been running, the
Pickup might be hot if it has been
lying near the exhaust manifold.

INDUCTIVE PICKUP

Fig. 3-13
Measuring rpm with the model 88 and an inductive pickup.

MAX Recording function also helps. Pressing MIN MAX enters MIN MAX Recording. Pressing MIN MAX three times steps to average (AVG).

Measuring RPM

When you measure rpm (Revolutions Per Minute) on the Fluke Model 88 DMM there
are two rpm functions available. The following procedure uses both such functions and the
inductive pickup (see Figure 3-13):

1. Turn engine off.
2. Connect output plug of Inductive Pickup to the input terminals (see Fig. 3-13).
3. Turn rotary switch to V DC.

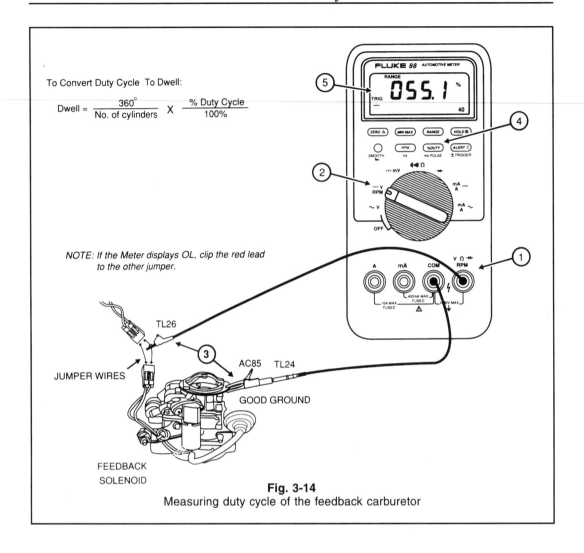

To Convert Duty Cycle To Dwell:

$$Dwell = \frac{360°}{No.\ of\ cylinders} \times \frac{\%\ Duty\ Cycle}{100\%}$$

NOTE: If the Meter displays OL, clip the red lead to the other jumper.

TL26

JUMPER WIRES

AC85   TL24

GOOD GROUND

FEEDBACK
SOLENOID

Fig. 3-14
Measuring duty cycle of the feedback carburetor

4. Press **RPM** to select RPM 2 for conventional 4-cycle engines (1 count/2 revolutions); RPM (2) appears on the DMM display. Press again to select RPM 1 for 2-cycle engines or waste spark 4-cycle engines (1 count/revolution); RPM (1) appears on the display.
5. Clamp the Inductive Pickup to a spark plug wire near the plug, see Fig. 3-13.
6. Turn the engine on. An rpm reading appears on the display. The engine should be turned off before removing the pickup.

This DMM can also take an rpm reading directly from the camshaft position sensor or tachometer by using test leads instead of the inductive pickup. When making an rpm measurement you may encounter the last digit on the display jumping around with various values. You can curb this by using the **SMOOTH** function.

Duty Cycle on a Feedback Carburetor

Duty cycle is the percentage of time (0 to 99.9 percent) a voltage is positive versus negative, or at ground potential. Most cars have their points closed for a duty cycle ranging from 50 to 70 percent. It is important to realize that dwell is the number of degrees of distributor rotation that the points remain open. Charts are available for converting duty cycle to dwell or you can use the formula in Fig. 3-14. The DMM can be set up in three easy steps:

1. Press % DUTY to select Duty Cycle; % is displayed.
2. Press this again to select Pulse Width Mode; ms+ or ms– is displayed.
3. Press this again when you wish to exit this mode.

In Duty Cycle (and Pulse Width), press **ALERT** (± trigger) to toggle between a positive and negative trigger slope. The slope is indicated by a + or a – sign below TRIG in the lower right-hand corner of the DMM's display.

The following test procedure measures duty cycle on a feedback carburetor (see Fig. 3-14 again).

1. Insert the test leads in the input terminals.
2. Set the rotary selector switch to DC volts.
3. Connect the test leads.
4. Press % DUTY. The DMM defaults to – trigger, and the display shows TRIG, –, and %.
5. Turn the car on and read the display when the engine is cold (open loop). Read the display again when the engine is warm (closed loop).

When your engine is in open-loop operation, the duty cycle is a steady value (consult the car's specifications). When a car's engine warms and goes into closed-loop operation, the reading should change up and down.

Pulse Width Measurements

The following is a handy measurement when troubleshooting a port fuel injector (see Fig. 3-15):

1. Connect the test leads as shown in the figure.
2. Set the rotary selector switch to DC volts.
3. Insert the test leads in the input terminals.
4. Press % DUTY twice to select Pulse Width.
5. The display shows ms.
6. Start the engine and read the DMM's display.

In the Pulse Width (and Duty Cycle) modes, the model 88 DMM defaults to (–) trigger slope (when the time signal is slow). Press **ALERT** (±**TRIGGER**) to toggle between slopes. The slope is indicated by the + or – sign below TRIG in the lower left-hand corner of the DMM's display.

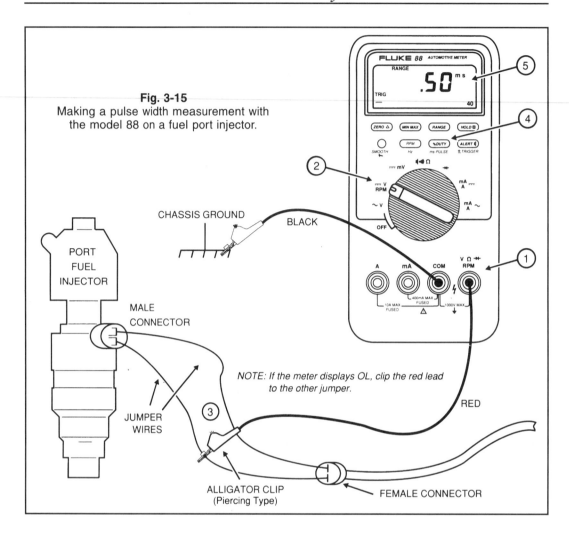

Fig. 3-15
Making a pulse width measurement with
the model 88 on a fuel port injector.

NOTE: If the meter displays OL, clip the red lead
to the other jumper.

Measuring Starter Circuit Voltage Drops

This troubleshooting procedure uses the Touch Hold™ feature of the Fluke Model 88 DMM. This function, though, does have its limitations. It will not allow you to faithfully capture unstable or noisy readings. Therefore, it shouldn't be used to determine if circuits with dangerous voltages are present or dead. If you were to manually change the Model 88 DMM after selecting Touch Hold™, the DMM exits the Touch Hold™ function. To select the Touch Hold™ function, you press **HOLD** H. A reverse video "H" appears on the display. Touch Hold™ operates in two ways. This depends on the function the meter is in when you select Touch Hold™. If you are in MIN MAX Recording, RPM, Duty Cycle, Pulse Width, or Hz when you select Touch Hold™, Touch Hold™ interrupts these functions. The DMM's display does not update, but the recorded readings are not erased. Pressing MIN MAX or SMOOTH when you are in Touch Hold™ causes the DMM to exit Touch Hold™ and enter MIN MAX

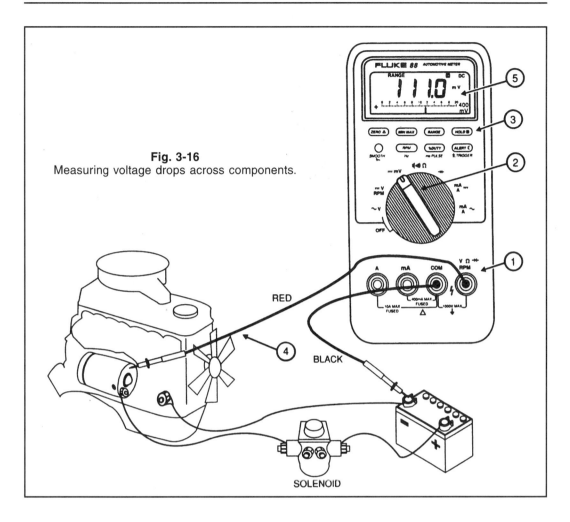

Fig. 3-16
Measuring voltage drops across components.

Recording or SMOOTH, respectively. If you are in MIN MAX Recording, RPM, Duty Cycle, Pulse Width, or Hz when you select Touch Hold™, the last stable reading is held on the display. When a new stable reading is detected, the beeper sounds and the display updates. This function allows you to take measurements in dangerous or difficult circumstances when you can't look at the display. You press **HOLD** H to exit Touch Hold™ and resume recording or counting. Figure 3-16 shows how to use this DMM feature to effectively measure voltage drops across components (except solenoids) and connections while you are inside your car cranking the engine. The procedure follows; note that your DMM is set to the millivolt range.

1. Insert the test leads in the input terminals (see Fig. 3-16 again.)
2. Set the rotary switch to mV DC.
3. Press **HOLD** to select Touch Hold™.
4. Touch the probes across the connection to be measured.
5. Crank your engine for 4-5 seconds. The DMM holds the voltage drop, referenced to ground, on its display.

Compensating for Lead Resistance

Earlier we mentioned how low of a resistance this special automotive DMM can measure. This resistance is at least 10 to 100 times less than the very resistance inherent in the test leads themselves. Therefore, you have to very carefully and accurately "zero" these test leads out before you can place any credibility in these very low resistances in the 10-m**W** range. The ZERO function on the model 88 DMM subtracts a stored value from the present measurement and displays the results. For example, if the stored measurement is 12.00 V and the present measurement is 4.10 V, the display shows –07.90 V. If the new reading is the same as the stored value, the display shows 0. When you press ZERO **A**, autoranging turns off, and you are locked into the range you are in. Use Manual Range (RANGE) to select the range before selecting ZERO. If you manually change the measurement range after you select ZERO, the DMM exits the ZERO function. The procedure for using ZERO **D** is as shown in Fig. 3-17.

1. Press ZERO **D** to select the ZERO function. The present reading is stored, the display reads 0, and **D** is displayed.

2. Press the button again (or change the switch setting) to clear the stored reading and exit the ZERO function.

3. Autoranging does not automatically resume when you exit the ZERO function. To exit Manual Range, press RANGE for more than two seconds.

Resistance in standard test leads greatly diminishes the accuracy of resistance measurements by 0.1 to 0.2 **W**. The fol-

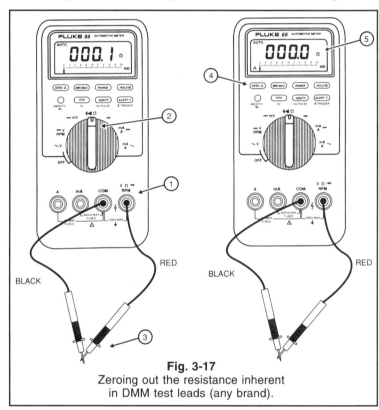

Fig. 3-17
Zeroing out the resistance inherent
in DMM test leads (any brand).

lowing test procedure explains how to use ZERO **D** to subtract test lead resistance from resistance measurements (see Fig. 3-17). When you take a reading, the test lead resistance is subtracted from the measurement before it is displayed:

1. Plug the test leads into the input terminals.
2. Set the rotary selector switch to **W**.

3. Touch the probes together. The display will typically read from 000.1 to 000.2 **W**.
4. While the test leads are shorted, press ZERO **D**.
5. The display shows 000.0 and **D**, indicating that ZERO is selected.

Selecting Power-Up Options

As previously mentioned, there are some DMM selections only possible when you first turn on the Model 88 automotive DMM. These appear on the back of the DMM (see Fig. 3-1 again). To select power-up options, press and hold down one or more pushbuttons while turning the rotary switch to any ON position. Power-up options remain selected until you turn off the DMM.

Checking a Car's Barometric Pressure / Manifold Absolute Pressure Sensors (BP/MAP)

There is a whole chapter virtually dedicated to your car's sensors, but for now, let's just review a little background. Figure 3-18 is the setup you use to make measurements and troubleshoot your BP/MAP sensor. This oxygen sensor, by its very defined use, has to be small, tough, and withstand a great variance in operating and outside environmental conditions. These include cold starts, hills, and stop-and-go traffic. On cars of the mid-1990s, airflow was measured and inferred mathematically. What that means is that by measuring key variables in your car's manifold, after sensing these key parameters, especially manifold air pressure (MAP), air temperature, engine rpm, and the mass of air entering the cylinder, a calculation is made by your car's on-board computer. However, there are inaccuracies inherent in this process due to the amount of time it takes to make all these slightly delayed calculations. As you vary your driving conditions, this inaccuracy is more pronounced. Naturally, as you cruise down the highway, on a level grade, and at a constant speed, these errors are greatly minimized.

One type of BP/MAP sensor works on the principle that as air passes over a heated element, the amount of heat it takes to maintain this element's heat constant is a key factor. As an example, Nichrome is a typical type of wire, which, when heated, has its resistance increased. This exhibits a positive temperature coefficient (PTC). Such an arrangement often consists of a quartz bobbin and ceramic bobbin. As air passes over this combination, it evacuates or "pulls" heat away from it. This is the same principle of operation as a downhill skier whose moist lips feel the added heat taken away because they are moist. In our example, the difference in temperature is a direct analogy between the current in the two elements. These "anemometers" are already in use in several luxury European cars, such as BMW. Bosch and Siemens use these types of sensors in their fuel injection systems. But their cost to the manufacturer of $30 makes them largely impractical until that price can be cut in half or more.

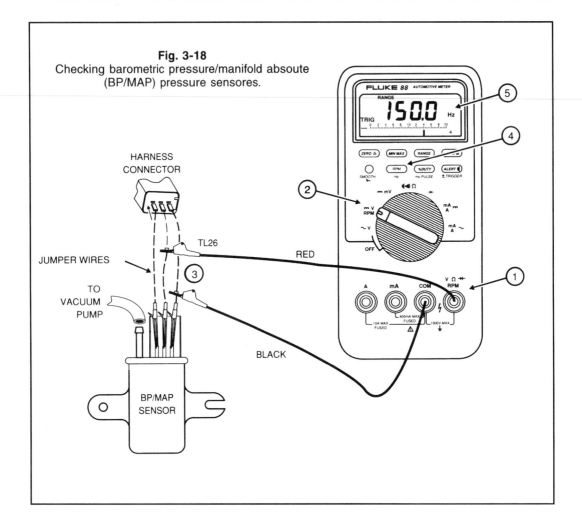

Fig. 3-18
Checking barometric pressure/manifold absoute
(BP/MAP) pressure sensores.

Measuring a Car's Starter Circuit Current Draw

Figure 3-19 shows you the setup to make a measurement of the current drawn at the starting of your car (current required to turn over your engine). This will help you isolate the source of starting problems. Figure 3-20 is a current transformer and these are often used when great amounts of current exist. They essentially provide an alternate path for current that the DMM knows and makes calculations accordingly. There are two types of current transformers. One is a clamp-on and the other requires you to thread the current-carrying conductor through it. This type can be a hassle, especially if you have to disconnect one end of the current-carrying wire or wires.

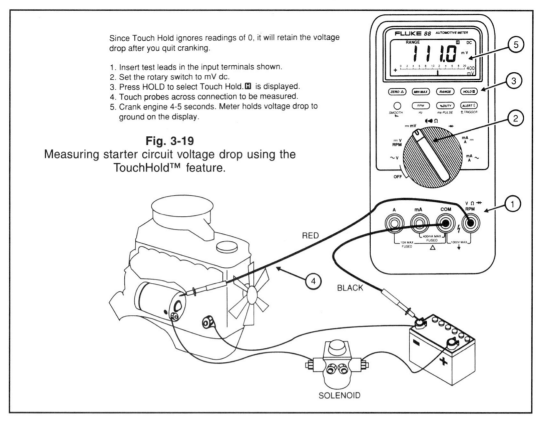

Since Touch Hold ignores readings of 0, it will retain the voltage drop after you quit cranking.

1. Insert test leads in the input terminals shown.
2. Set the rotary switch to mV dc.
3. Press HOLD to select Touch Hold. ⊞ is displayed.
4. Touch probes across connection to be measured.
5. Crank engine 4-5 seconds. Meter holds voltage drop to ground on the display.

Fig. 3-19
Measuring starter circuit voltage drop using the TouchHold™ feature.

Fig. 3-20
A clamp-on type current transformer

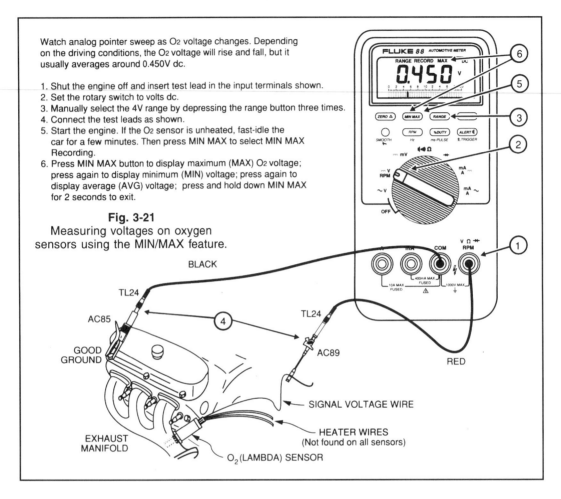

Watch analog pointer sweep as O2 voltage changes. Depending on the driving conditions, the O2 voltage will rise and fall, but it usually averages around 0.450V dc.

1. Shut the engine off and insert test lead in the input terminals shown.
2. Set the rotary switch to volts dc.
3. Manually select the 4V range by depressing the range button three times.
4. Connect the test leads as shown.
5. Start the engine. If the O2 sensor is unheated, fast-idle the car for a few minutes. Then press MIN MAX to select MIN MAX Recording.
6. Press MIN MAX button to display maximum (MAX) O2 voltage; press again to display minimum (MIN) voltage; press again to display average (AVG) voltage; press and hold down MIN MAX for 2 seconds to exit.

Fig. 3-21
Measuring voltages on oxygen sensors using the MIN/MAX feature.

Measuring a Car's Oxygen Sensor Voltage

Figure 3-21 is a setup showing how to make measurements on your oxygen sensor. As background, your oxygen sensor sends its voltage to the ECM (computer). This controls fuel injector on-time, or carburetor dwell (its duty cycle of what proportion of a cycle it remains on). This is naturally to maintain the ideal air-to-fuel mixture of 14.7:1. Zirconia oxygen sensors act as batteries. They generate a voltage output proportional to the amount of oxygen they sense. Therefore, you find many Zirconia sensors used in many places where output is in the form of resistance that changes in response to oxygen level changes. In more definite terms, if the amount of time oxygen is present equals the time there is no oxygen, its output voltage is 500 mV or one-half volt. You can deduce that you have a lean condition in your fuel injection system if this voltage is less than 500 mV. You have a rich condition if it is above 500 mV. By design, these sensors oscillate back and forth between "oxygen" and "no oxygen" and span the voltage range of 200 mV to 800 mV. If your sensor is outside this range, change it.

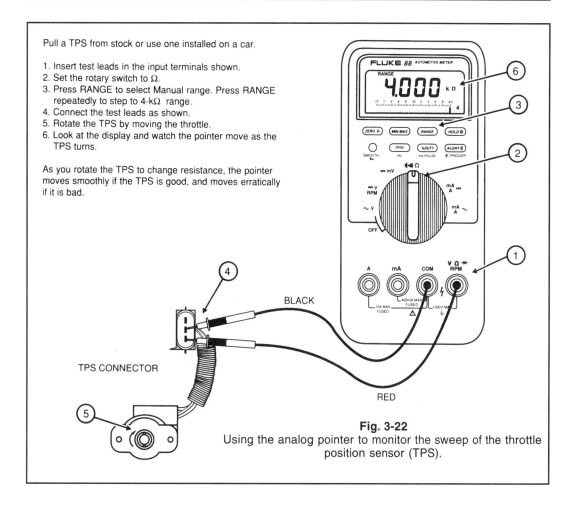

Pull a TPS from stock or use one installed on a car.

1. Insert test leads in the input terminals shown.
2. Set the rotary switch to Ω.
3. Press RANGE to select Manual range. Press RANGE repeatedly to step to 4-kΩ range.
4. Connect the test leads as shown.
5. Rotate the TPS by moving the throttle.
6. Look at the display and watch the pointer move as the TPS turns.

As you rotate the TPS to change resistance, the pointer moves smoothly if the TPS is good, and moves erratically if it is bad.

Fig. 3-22
Using the analog pointer to monitor the sweep of the throttle position sensor (TPS).

Measuring Your Car's Throttle Position Sensor

Figure 3-22 shows the setup to accomplish this, but first, a little theory. Every electronic fuel injection system uses a throttle position sensor (TPS). It supplies the engine control microprocessor with a signal proportionate to the opening angle of the throttle plate(s). It is generally a potentiometer or variable resistor. If it is adjustable, it will have slotted mounting screw holes. These allow the sensor to be repositioned until a specified voltage (usually 1 volt or less), results when your throttle is in the closed position.

To troubleshoot this you would:

1. Connect your DMM to a breakout box, such as the Ford EEC-IV.
2. Connect the + lead of your DMM to pin 47 and the – lead to pin 46 of this device's connector.
3. Turn the ignition switch on, but do not start your engine yet.

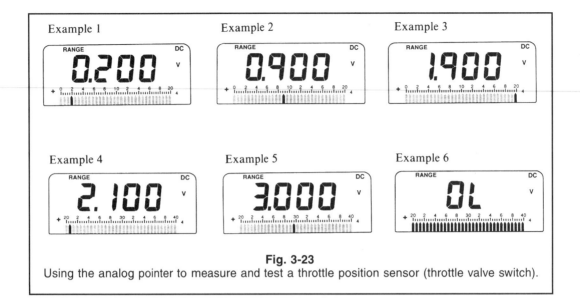

Fig. 3-23
Using the analog pointer to measure and test a throttle position sensor (throttle valve switch).

Using the DMM's Analog Pointer Feature to Test Your Car's Throttle Valve Switch

Figure 3-23, the analog pointer feature of the Model 88 DMM acts very much like the needle does on an analog DMM. You read the analog display when a rapidly changing signal makes the digital display hard to read. As an example, the pointer is fast enough that it captures contact bounce when relays close. In most functions, the pointer moves across the DMM's scale two times per each range. If you discover that the pointer is too sensitive, move to a higher scale. This pointer feature is disabled during the following modes: rpm, % Duty Cycle and ms-Pulse Width. The six illustrations within Figure 3-23 show the equivalent of an analog display. In the 4V range, the first time across the scale, the digital and analog display shows examples 1-3. The second time across the display's scale, the digital and analog display shows examples 4-6.

Chapter 3 Quiz

True or False

1. Fluke manufactures the model 88 DMM.
2. This DMM has eight pushbuttons.
3. This DMM does not allow you to change your measurement resolution.
4. This DMM's display is solely digital in nature.
5. A car's open circuit battery voltage test is performed with all of the car's accessories on.
6. A continuity check means that you verify a path exists whereby current can travel through a conductor, back to its source, thus completing a circuit.
7. The term duty cycle refers to the proportion of time a device is active (ON).
8. You often make pulse width measurements on your fuel port injectors.
9. The BP/MAP sensor's primary function is to measure humidity.
10. A clamp-on ammeter enables you to measure very minute currents that otherwise would be impossible to measure without this handy accessory to your DMM.

Fill in the Blanks

11. The 8 position rotary switch on the Fluke model 88 DMM selects _____ functions.
12. The Fluke model 88 DMM displays each element that could possibly be active in its display upon the _____ upon first powering up.
13. When you know in advance that you have a continuously varying parameter to measure, such as voltage, a handy feature to use is the _____.
14. A _____ diode has a 0.3 V voltage drop across it.
15. When attempting to determine the source of an unwanted current drain on your car's battery, you need to make certain that your battery has not been recently _____.
16. You use the _____ scale when making measurements on your ignition coils.
17. If you experience a lot of electrical noise that makes your readings rapidly change, you can solve this by changing to a _____ scale.
18. When you are making RPM measurements, you need to use an _____.
19. Most cars have their _____ close for duty cycles from 50 to 70 percent.
20. When you want to determine the duty cycle of your port fuel injector, you would first find the wave form or signal _____.

4

Getting Acquainted with a DSO and ScopeMeter®

Despite this chapter's lengthy title, it still fails to fully explain what we are attempting to learn here. It omits one key ingredient: that you will first have to learn how an ordinary analog oscilloscope (scope) works before moving onto DSOs (Digital Storage Oscilloscopes). If the theory presented in this chapter agrees with you, then there will be few, if any, signals your car generates that you can't capture and properly analyze. Admittedly, the DSOs we will use can perform extremely sophisticated measurements, far beyond what many cars presently require. How? Be patient! Modern cars are becoming more complex, especially with the number of sensors they now possess. There is a California law going into effect soon requiring a sensor in the transmission to have it shift better, and in turn give you better fuel economy. This book, in that sense at least, won't be obsolete in two years as so many computer books are.

A scope is one of the most useful pieces of electronic test equipment you will ever use. We'll begin with an introduction to the display, horizontal, and vertical systems of the instrument. We'll also make some fundamental observations and definitions of various types of waveforms. We will then progress to the scope's triggering system, its probes, and show you how to get started using the scope. We will then look at actual applications using a scope to make critically important measurements. Figure 4-1 is a typical dual-trace oscilloscope.

A Scope's System and Controls

A Scope's Function

Scopes analyze electrical signals. However, these electrical signals may have been converted from another source of energy. This is done by using transducers, devices that change energy from one form to another. Therefore, they are very useful in measuring nonelectrical phenomena. Speakers and microphones are examples of transducers. Speakers convert electri-

Fig. 4-1
A dual trace scope that is unsuitable for car troubleshooting in a garage for three reasons: 1) It is too delicate; if it drops, the CRT will shatter and the unit breaks; 2) It does not work off a battery; 3) It is too large and bulky to fit in small places.

cal energy into mechanical energy. Conversely, microphones convert mechanical energy (supplied by your voice) into electrical signals. Other transducers convert heat, mechanical stress, and light into electrical signals, all of which can be seen and analyzed on a scope.

Overview Block Diagram

Figure 4-2 is a block diagram of an oscilloscope. As an overview, the vertical system controls the vertical axis of the graph. When the electron beam draws a line or trace in a vertical direction, it does so under the control of this system. The horizontal system controls the left-to-right movement of the beam. The trigger system determines when the scope begins drawing these horizontal sweeping lines. The actual display system primarily consists of a CRT (Cathode Ray Tube) and its drive circuitry. It is this tube where all traces are drawn.

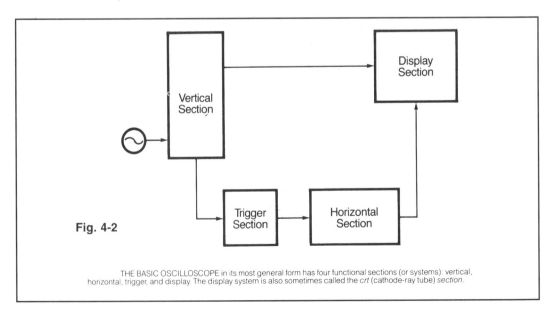

Fig. 4-2

THE BASIC OSCILLOSCOPE in its most general form has four functional sections (or systems): vertical, horizontal, trigger, and display. The display system is also sometimes called the *crt* (cathode-ray tube) *section*.

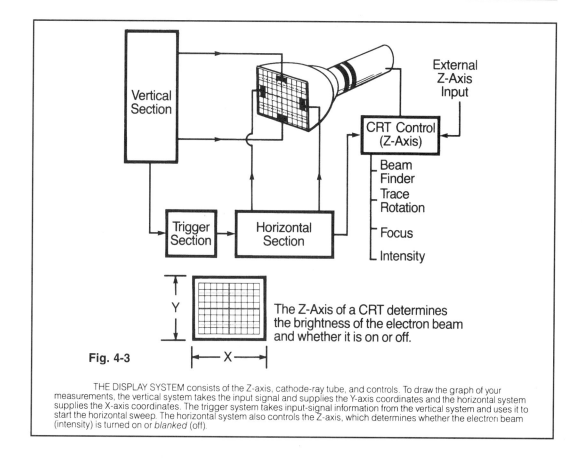

Fig. 4-3

The Z-Axis of a CRT determines the brightness of the electron beam and whether it is on or off.

THE DISPLAY SYSTEM consists of the Z-axis, cathode-ray tube, and controls. To draw the graph of your measurements, the vertical system takes the input signal and supplies the Y-axis coordinates and the horizontal system supplies the X-axis coordinates. The trigger system takes input-signal information from the vertical system and uses it to start the horizontal sweep. The horizontal system also controls the Z-axis, which determines whether the electron beam (intensity) is turned on or *blanked* (off).

The *Display System*, as previously stated, draws a graph by moving an electron beam across a phosphor coating on the inside of the CRT (see Fig. 4-3). The result is a glow, which, for a short time afterward, traces the path of the beam. There is a grid of lines etched or silk-screened on the inside of the CRT faceplate. This is called the graticule. These serve as the reference for measurements, see Fig. 4-4.

The *Intensity Control* adjusts brightness of the trace. Adjustment of intensity is necessary not solely because of different ambient conditions. Square waves, as an example, have slower horizontal segments than their faster vertical segments. As such, you would increase the intensity to see the fainter vertical segments.

The *Beam Finder* lets you locate the electron beam when it is off the screen. When you press this button you are actually reducing the horizontal and vertical amplifier voltages (more about this later). This also overrides the INTENSITY control. The trace can then appear within the 80 x 100-millimeter screen. After the beam appears in one of the screen's four quadrants, you will then know how to adjust the HORIZONTAL and VERTICAL position controls.

The *Focus Control* within a scope is controlled by a signal on a grid within the CRT. This ensures that the electron beam is focused on the CRT faceplate.

Trace Rotation allows you to electrically align the horizontal deflection of the trace with the fixed CRT graticule. To avoid accidental misalignments, it is usually recessed or kept out of the way.

The *Vertical System* Figure 4-5 shows how this system supplies the display system with the vertical or Y-axis data. These input signals deflect the CRT beam. The vertical system also gives you a choice over input coupling modes: AC or DC. Some scopes have a TRACE SEP (Separation) control and three mode switches. These allow you to select Channels 1 or 2 (on a dual-channel scope), or both channels, invert one channel, algebraically add two channels, alternately display, or chop the display.

Coupling Controls. Vertical channel input coupling controls allow you to select either AC or DC input coupling and ground (see Fig. 4-6). Note in this that the photo on top has ground referenced to midway up the second of 10 vertical divisions. The other photo has ground referenced right in the middle of the screen. DC coupling connects the entire input signal to the vertical amplifier. AC coupling blocks DC voltages and only allows alternating current components to pass. The GND (ground) position disconnects the input signal and shows you the scope's true chassis ground.

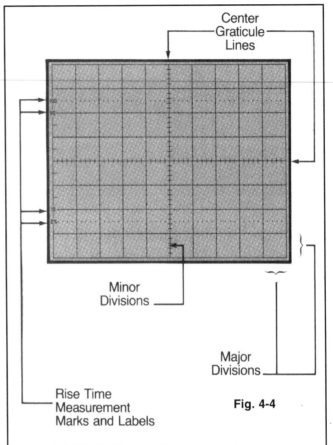

Fig. 4-4

THE GRATICULE is a grid of lines typically etched or silk-screened on the inside of the crt faceplate. Putting the graticule inside—on the same plane as the trace drawn by the electron beam—eliminates measurement inaccuracies called *parallax errors*. Parallax error occurs when the trace and the graticule are on different planes and the observer shifts slightly from the direct line of sight. Though different-sized crt's may be used, graticules are usually laid out in an 8-by-10 pattern. Each of the 11 vertical and 9 horizontal lines block off major divisions (also simply called *divisions*) of the screen. Labeling on the scope controls always refers to major divisions. The tick marks on each of the graticule lines represent minor divisions or subdivisions Since scopes are often used for rise time measurements, 2200-Series scope graticules include special markings to aid in making rise-time measurements There are dashed lines for 0% and 100% levels and labeled graticule lines for the 10% and 90% points (where rise time is measured).

AC coupling is handy when the entire signal (alternating plus constant components) might be too large for the VOLTS/DIV switch setting. In this case, you might see something like the top photo, but eliminating the DC allows you to see the AC component (shown in the bottom photo).

Input Connectors. Almost every scope uses BNC connectors as input "avenues". Historically, a BNC derived its name from its two inventors: The *B* is for Bayonet type connector

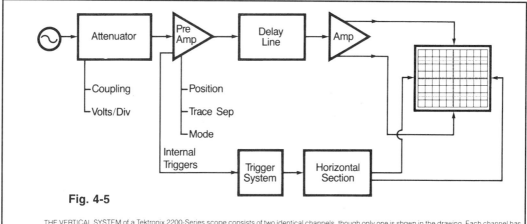

Fig. 4-5

THE VERTICAL SYSTEM of a Tektronix 2200-Series scope consists of two identical channels, though only one is shown in the drawing. Each channel has circuitry to couple an input signal to that channel, attenuate the signal (that is, reduce it) preamplify it, delay it, and finally amplify the signal for use by the display system. The delay line lets you see the beginning (or leading edge) of a waveform even when the scope is triggering on it.

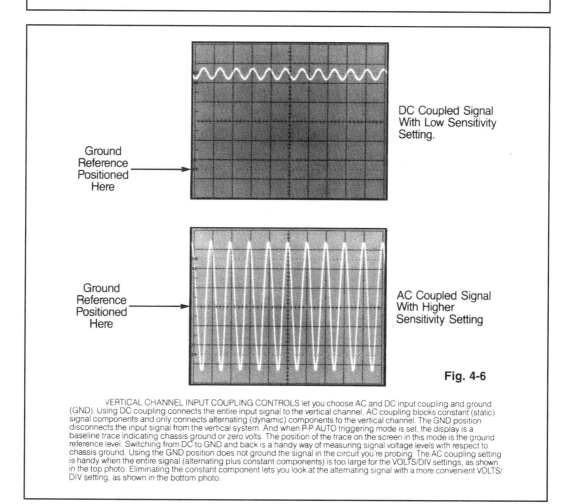

Ground Reference Positioned Here

DC Coupled Signal With Low Sensitivity Setting.

Ground Reference Positioned Here

AC Coupled Signal With Higher Sensitivity Setting

Fig. 4-6

VERTICAL CHANNEL INPUT COUPLING CONTROLS let you choose AC and DC input coupling and ground (GND). Using DC coupling connects the entire input signal to the vertical channel. AC coupling blocks constant (static) signal components and only connects alternating (dynamic) components to the vertical channel. The GND position disconnects the input signal from the vertical system. And when P-P AUTO triggering mode is set, the display is a baseline trace indicating chassis ground or zero volts. The position of the trace on the screen in this mode is the ground reference level. Switching from DC to GND and back is a handy way of measuring signal voltage levels with respect to chassis ground. Using the GND position does not ground the signal in the circuit you're probing. The AC coupling setting is handy when the entire signal (alternating plus constant components) is too large for the VOLTS/DIV settings, as shown in the top photo. Eliminating the constant component lets you look at the alternating signal with a more convenient VOLTS/DIV setting, as shown in the bottom photo.

with the *N* derived from Paul Neill and the *C* from Carl Concelman, who developed the C-series connectors while working at Bell Labs.

The *Vertical Position Control* changes the trace's position on the CRT to the desired position.

Vertical Sensitivity is controlled by the VOLTS/DIV switch. Typical scopes go down to 2 mV/div. But remember, when examining larger signals you may want to attenuate the input signal with a 10:1 probe. Vertical magnification also exists on some scopes. This is in the form of a pull-out knob on the VOLTS/DIV switch. This knob is sometimes marked "CAL" and increases sensitivity by a factor of 10:1. However, this reduces the scope's bandwidth to typically 5 MHz or less. If you do not pull out this CAL center knob within the VOLTS/DIV switch, you can use it to change the scale factor of the VOLTS/DIV setting by a factor of greater than 2.5 times in either direction.

Vertical Operating Modes. As previously stated, you may select Channel 1 alone, Channel 2 alone, both channels in either alternate or chopped modes or both channels algebraically added. As an example, to see both channels in the alternate mode, set one switch to BOTH and the other vertical mode switch to ALT. This causes the scope to draw each trace alternately. To display both channels in the chopped mode, set the switch which was previously in the ALT position to the CHOP position. This draws small parts of each waveform by switching back and forth between them. The switching is so fast that it tricks your eye into mentally filling in the missing gaps. This mode is typically used in applications with sweep speeds of 1 msec. or slower.

Both chopped and alternated modes are provided so you can observe two signals at any sweep speed. The ALT mode first draws one signal, then the next, but not at the same speed. This works well at faster sweep speeds, when your eye can't detect the alternating actions. If you wanted to see two signals (simultaneously) at slower sweep speeds, you'd use the CHOP mode.

To see two distinct and different signals displayed as one waveform, you'd use the BOTH and ADD modes. This algebraically combines (adds) channels 1 and 2. But if you prefer, it can also subtract channel 2 from channel 1 when channel 2 is in the INVERT mode.

The *Horizontal System.* Figure 4-7 shows how this system supplies the deflection voltages which move the electron beam horizontally. It also contains a sweep generator which produces a voltage ramp, see Fig. 4-8. This ramp controls the scope's sweep rates. The importance of the sweep generator is it produces a linear rate of rise in the ramp. This makes it possible to calibrate the movement of the horizontal beam in units of time; therefore, it is usually called the time base in a modern scope. This time base allows you to observe the signal over any unit of time you select. This can vary from as little as a few nanoseconds, up to and including a whole second.

Horizontal Operating Modes

Single time base scopes, and the vast majority of scopes are single time base scopes, let you select one horizontal operating mode. However, there are dual time base scopes which

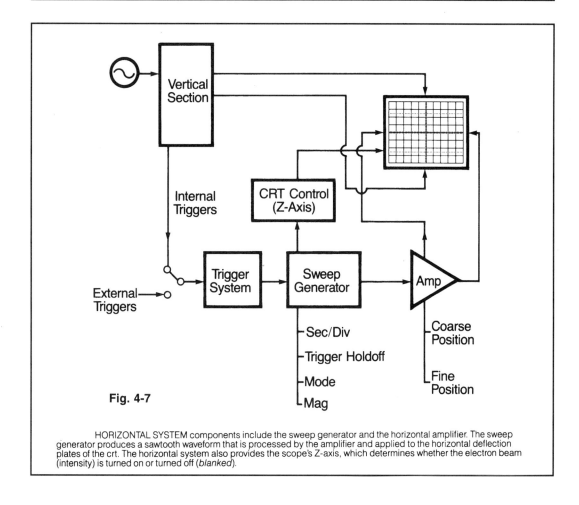

Fig. 4-7

HORIZONTAL SYSTEM components include the sweep generator and the horizontal amplifier. The sweep generator produces a sawtooth waveform that is processed by the amplifier and applied to the horizontal deflection plates of the crt. The horizontal system also provides the scope's Z-axis, which determines whether the electron beam (intensity) is turned on or turned off (*blanked*).

allow you to select either of two sweeps. The A sweep can remain undelayed, while the B sweep can be delayed to start after a delay you select.

Sweep Speeds

The second/division switch lets you select sweep speed, which is, the rate at which the beam sweeps across the screen. Like the vertical system's VOLTS/DIV switch, this control's markings refer to the screen's scale factor.

Variable Sweep Speeds

Besides the calibrated sweep speeds, some scopes have a control in the center of the circular sweep speed dial which is often red. This dial, when rotated counterclockwise, slows the sweep speed by at least a 2.5:1 ratio.

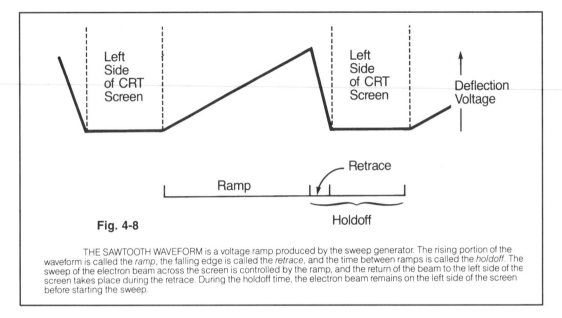

Fig. 4-8

THE SAWTOOTH WAVEFORM is a voltage ramp produced by the sweep generator. The rising portion of the waveform is called the *ramp*, the falling edge is called the *retrace*, and the time between ramps is called the *holdoff*. The sweep of the electron beam across the screen is controlled by the ramp, and the return of the beam to the left side of the screen takes place during the retrace. During the holdoff time, the electron beam remains on the left side of the screen before starting the sweep.

Horizontal Magnification

Most scopes have some means of horizontally magnifying the waveforms. This is done by actually multiplying the sweep speed by the amount of magnification. On a 10 X magnifier, this would be 10 times. For example, a 5 usec/division sweep speed would be magnified to a faster 0.5 usec/division. We have now examined how scopes write on the screen and the controls enabling you to do this. Let's conclude by defining several types of waveforms.

Waves

The definition of a wave is a disturbance traveling through a medium. A waveform is a graphic representation of this wave, which is just what a scope displays.

Like a wave, a waveform depends on both movement and time like the ripples from a pebble thrown into a pond. The waveform on your scope is the movement of an electron beam during time.

Basic Waveforms

Figure 4-9 shows a sine wave and various nonsine wave or nonsinusoidal waves such as a square wave, a triangle wave and a sawtooth wave. A square wave has two equal amounts of time for its two states. Triangle and sawtooth waves are usually the result of circuits designed to control voltage with respect to time. Typical of these are the sweep of a scope and certain circuits with a TV. In these type circuits, changes from one transition to another are made with a steady variation at a constant rate, which is called a ramp. Changes from one state to another

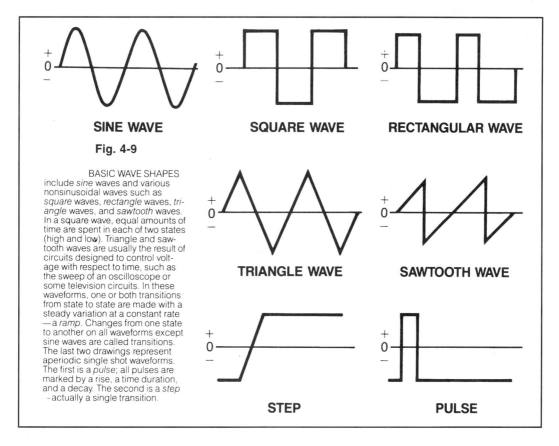

SINE WAVE

SQUARE WAVE

RECTANGULAR WAVE

Fig. 4-9

BASIC WAVE SHAPES include *sine* waves and various nonsinusoidal waves such as *square* waves, *rectangle* waves, *triangle* waves, and *sawtooth* waves. In a square wave, equal amounts of time are spent in each of two states (high and low). Triangle and sawtooth waves are usually the result of circuits designed to control voltage with respect to time, such as the sweep of an oscilloscope or some television circuits. In these waveforms, one or both transitions from state to state are made with a steady variation at a constant rate —a *ramp*. Changes from one state to another on all waveforms except sine waves are called transitions. The last two drawings represent aperiodic single shot waveforms. The first is a *pulse*; all pulses are marked by a rise, a time duration, and a decay. The second is a *step* –actually a single transition.

TRIANGLE WAVE

SAWTOOTH WAVE

STEP

PULSE

on all waveforms, with the exception of a sine wave, are called transitions. The last two waveforms, representing a periodic or single occurrence waveform, do not repeat themselves. The first waveform is a pulse with a rise, a definite duration, and a decay or fall time. The second waveform is a step and has no decay until the circuit is de-energized.

We have just seen how a scope manages to draw waveforms of various types onto its CRT. Now, let's look at triggering a scope and how to use it safely. We'll also see how a simple probe affects a waveform and we'll start using the scope by making some very basic measurements.

The Trigger System

Now we know how the scope draws a waveform, but we don't know when this happens. Let's realize measuring time related information is one of the main reasons for using a scope. Therefore, it is crucial to control and measure when an event happens and for how long. As an example, if you are using the 0.5 usec/division setting and drawing one graph every 5 usecs or 0.5 usec/division times 10 screen divisions, you are actually seeing 200,000 traces every second (not counting retrace and hold-off times). Imagine how chaotic the CRT would be if every trace started at a different time, the screen would be a total blur or continuous smudge. But

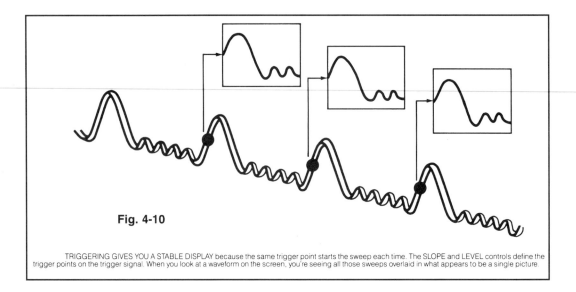

Fig. 4-10

TRIGGERING GIVES YOU A STABLE DISPLAY because the same trigger point starts the sweep each time. The SLOPE and LEVEL controls define the trigger points on the trigger signal. When you look at a waveform on the screen, you're seeing all those sweeps overlaid in what appears to be a single picture.

each trace starts at the same time, after you make the right trigger control setting. How is all of this done? Simply connect the trigger signal to the trigger system circuit with the external coupling controls we've already discussed.

Next, set the trigger to recognize a certain voltage level and slope. Figure 4-10 illustrates this process. Note how each time the trigger occurs on the rising side of this waveform.

When you observe a properly triggered signal on the scope's CRT you are actually seeing all those sweeps overlaid into what appears is a single picture.

Trigger Level and Slope Controls

These controls define the trigger point. Note in Fig. 4-11 that the SLOPE control determines if a positive or negative going edge sets the trigger or starting point of the waveform. The LEVEL control sets where on the waveform (and at what voltage level) this takes place.

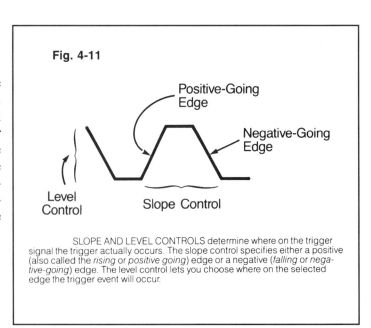

Fig. 4-11

Positive-Going Edge

Negative-Going Edge

Level Control

Slope Control

SLOPE AND LEVEL CONTROLS determine where on the trigger signal the trigger actually occurs. The slope control specifies either a positive (also called the *rising* or *positive going*) edge or a negative (*falling* or *negative-going*) edge. The level control lets you choose where on the selected edge the trigger event will occur.

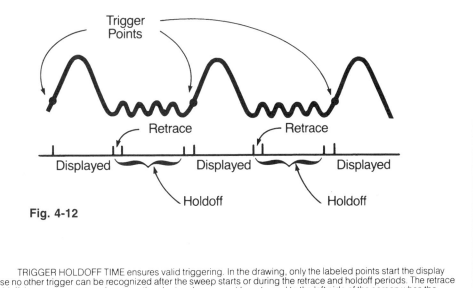

Fig. 4-12

TRIGGER HOLDOFF TIME ensures valid triggering. In the drawing, only the labeled points start the display because no other trigger can be recognized after the sweep starts or during the retrace and holdoff periods. The retrace and holdoff times are necessary because the electron beam must be returned to the left side of the screen when the sweep ends, and because the sweep generator needs the reset time. The crt Z-axis is *blanked* (or turned off) between sweeps and *unblanked* during sweeps.

Variable Trigger Hold-off

Not every trigger event is valid. As an example, during sweep or retrace, and for a short period after called "the holdout period," the trigger is invalid. Note in Fig. 4-12 the areas called "DISPLAYED". These are the only times allowing triggering.

Sometimes the normal hold-off period isn't long enough to make certain you get a stable display. This possibility exists when you are examining a particularly complex waveform such as a long digital streams which is periodic (repeats itself), but still has a number of trigger points on it which might otherwise cause false triggering. In these cases you are given a series of patterns rather than the same one each time since it triggers at various places within the waveform. Obviously, what you need is some means to control when a trigger is accepted. Figure 4-13 is a diagram of variable hold-off controls and note how the scope ignores some potential trigger points which might otherwise portray a confusing display of the "real" waveform.

Trigger Sources

These sources are grouped into two categories: external and internal. It makes no difference in how the trigger circuit operates, but internal triggering usually means the scope is also displaying the waveform it is triggering upon. There are usually two switches on a scope's front panel called SOURCE and INT which control this.

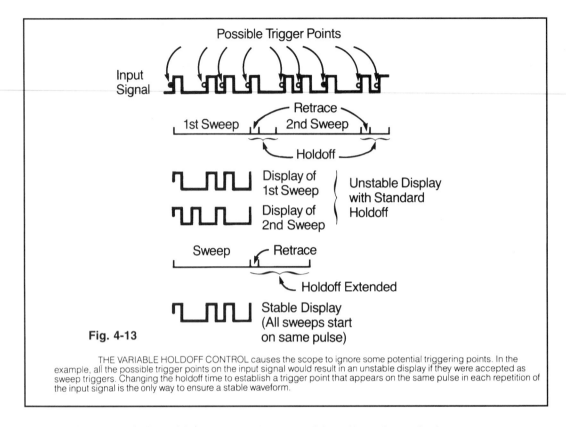

Fig. 4-13

THE VARIABLE HOLDOFF CONTROL causes the scope to ignore some potential triggering points. In the example, all the possible trigger points on the input signal would result in an unstable display if they were accepted as sweep triggers. Changing the holdoff time to establish a trigger point that appears on the same pulse in each repetition of the input signal is the only way to ensure a stable waveform.

In the INT switch position you can trigger on either channel on a dual trace scope or even switch to the VERT mode. In the VERT mode the internal source switch is used for triggering. If the VERT mode switch is set at channel 1, then signals on Channel 1 trigger the scope. If they are set on channel 2, that channel triggers the scope. If you switch to the alternate (ALT) vertical mode, the scope looks for signals alternately on the scope. If the VERT mode is in the ADD position, the scope adds the algebraic sum of channels 1 and 2.

Oscilloscope Applications

We were introduced to the oscilloscope, the most useful piece of overall test equipment on your bench. Now let's start making actual measurements and using it for more practical applications such as testing stereo equipment and checking electronic components such as resistors, capacitors and inductors.

Starting with the Basics

Since every measurement you will make involves either a time or amplitude measurement, let's review how to make these two fundamental measurements first.

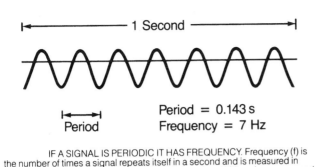

Period = 0.143 s
Frequency = 7 Hz

IF A SIGNAL IS PERIODIC IT HAS FREQUENCY. Frequency (f) is the number of times a signal repeats itself in a second and is measured in hertz: 1 Hz = 1 cycle per second, 1 kHz (kilohertz) = 1000 cycles per second, and 1 MHz (megahertz) = 1,000,000 cycles per second. Period and frequency are reciprocal: 1/period = frequency, and 1/frequency = period. For example, a 7-Hz signal has a period of 0.143 seconds, since: 1 cycle ÷ 7 Hz = 0.143 s and 1 cycle ÷ 0.143 s = 7 Hz.

Fig. 4-14

Amplitude Measurements

It is best to make amplitude measurements by filling up most of the screen vertically. This improves the resolution, and thus the accuracy of the measurement, see Fig. 4-14 and follow this procedure:

1. Connect the Channel 1 probe to the Channel 1 connector and attach the probe tip to the PROBE ADJ terminal. If it is a dual trace scope, attach the probe ground lead to the Channel 2 connector.

2. Set the VERTICAL MODE switch to Channel 1 and the Channel 1 input switch to AC. The TRIGGER MODE switch should be in NORM. Set TRIGGER SOURCE switch to Channel 1.

3. Obtain a stable trace by adjusting the TRIGGER LEVEL control and move the Channel 1 VOLTS/DIV switch to a position at which the probe adjust square wave is just about five divisions.

4. Use the Channel 1 POSITION control to move the square wave so that its bottom is aligned with a convenient horizontal graticule line. That allows you to very nearly center the waveform vertically.

5. Count the major and minor divisions up the center vertical graticule and multiply by the VOLTS/DIV. setting. As an example, if you were set at 0.1 volts/division, and you had 5.8 divisions, you would have a square wave with a .58 volt amplitude.

Time Measurements

Making time measurements (see Fig. 4-15) is best while using the horizontal graticule line. The setup procedure is the same as the time measurement example just completed. Then, with the horizontal coarse and fine controls, line up the rising edge of the square wave with the vertical graticule which is second from the left of the CRT screen. Count major and minor graticule divisions across the center horizontal graticule of the screen and multiply by the SEC./DIV. setting. As an example, if there were 7.2 divisions in this example and the SEC./DIV. setting was 1 ms (millisecond), then the time would be 7.2 milliseconds and the frequency would be the reciprocal of that or 1 / .0072 seconds or 138.88 Hz.

Derived Measurements are the result of calculations made after making a direct measurement. The past example was a derived measurement since we actually measured time, and from that calculated frequency.

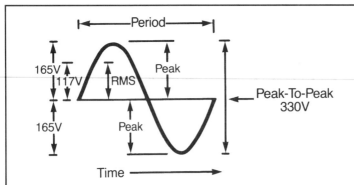

DERIVED MEASURE-MENTS are the result of calculations after making direct measurements. For example, alternating current measurements first require an amplitude measurement. The easiest place to start is with a peak-to-peak amplitude measurement of the voltage—in this case, 330 volts—because peak-to-peak measurements ignore positive and negative signs. The *peak voltage* is one-half that value (when there is no DC offset), and is also called *maximum value*; in this case it's 165 V. The *average value* is the total area under the voltage curves divided by the period in radians; in the case of a sine wave, the average value is

zero because the positive and negative values are equal. But in some applications such as power, the average value is determined to be:

$$V_{avg} = 0.318 \times V_{p\text{-}p}$$

The rms (root-mean-square) voltage for this sine wave—which represents the line voltage in the United States—is:

$$V_{rms} = \frac{V_p}{\sqrt{2}} = \frac{165}{1.414} = 117 \text{ V}$$

You get from peak-to-peak to rms voltage with:

$$V_{rms} = \frac{V_{p\text{-}p}}{\sqrt{2}}$$

Fig. 4-15

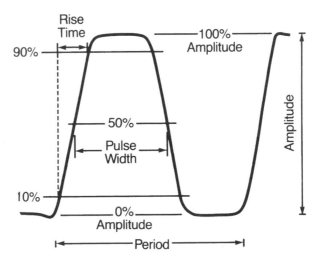

THE PARAMETERS OF A PULSE can be important in a number of different applications, including digital circuitry, X ray equipment, and data communications. Pulse specifications include transition times measured on the leading edge of a positive-going transition; this is the *rise time*. *Fall time* is the transition time on a negative-going trailing edge. *Pulse width* is measured at the 50% amplitude points, and *amplitude* is measured from 0% to 100%. Any displacement of the base of the pulse from zero volts is the *dc offset*.

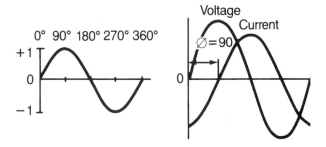

0° 90° 180° 270° 360°

PHASE IS BEST EXPLAINED WITH A SINE WAVE. Remember that this waveform type is based on the sine of all angles from 0° through 360°. The result is a plot that changes from 0 at 0°, 1 at 90°, 0 again at 180°, − 1 at 270°, and finally 0 again at 360°. Consequently, it is useful to refer to the *phase angle* (or simply *phase*) of a sine wave when you want to describe how much of the period has elapsed. In another usage, *phase shift* describes a relationship between two signals. Picture two clocks with their second hands sweeping the dial every 60 seconds. If the second hands reach twelve at the same time, the clocks are *in phase*; if they don't, the clocks are *out of phase*. *Phase shift* expresses the amount that two signals are out of phase. To illustrate, the waveform labeled current in the drawing is said to be 90° out of phase with the voltage waveform. Other ways of reporting the same information are: the Current waveform lags the Voltage waveform by 90°, or the Current waveform has a 90° phase angle with respect to the Voltage waveform. Note that there is always a reference to another waveform; in this case, the Current waveform relative to the Voltage waveform for an inductor.

Fig. 4-16

Note in Fig. 4-16 that this waveform has defined peak-to-peak, peak, rms (Root Mean Square), and average AC voltages. Refer again to Fig. 4-9 to see how these various waveforms relate to each other. As a refresher, rms value is also called the effective value of voltage. This is because it is the "effective" value of an alternating current, which must flow through a given resistance, to produce an equal heating effect in the same time interval.

Measuring Phase

Waveforms related by phase can be measured in two ways. First, the simplest way is to connect each waveform to the two vertical inputs of a dual trace scope. You can view them in either the CHOPPED or ALT VERT (alternated vertical) modes.

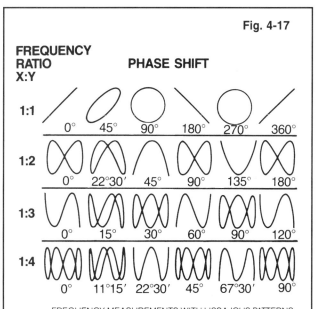

Fig. 4-17

FREQUENCY MEASUREMENTS WITH LISSAJOUS PATTERNS require a known sine wave on one channel. If there is no phase shift, the ratio between the known frequency (usually applied to the X input) and the unknown frequency (applied to the Y input) corresponds to the ratio between the number of vertical loops and the number of horizontal loops in the pattern. When the frequencies are the same, only the shifts in phase affect the pattern. In the drawings here, both phase and frequency differences are shown.

Measuring Phase with Lissajous Patterns

This is the second way of measuring phase. This method, named after the French physicist Jules Antoine Lissajous, relates shapes to phase angle difference. Lissajous measurements are limited with general purpose oscilloscopes to the bandwidth of the horizontal amplifier. This is usually much less than the vertical amplifier bandwidth. Figure 4-17 shows that a sine wave is required as an input to one channel of the scope.

If there is no phase shift, the ratio between the known frequency (usually applied to the X input) and the unknown frequency (usually applied to the Y input), corresponds to the between the number of vertical loops and the number of horizontal loops in the pattern. When the frequencies are the same, only the shifts in phase affect the pattern. Refer again to Fig. 4-17 and note how there is a 1:1 ratio in frequencies on the top line and a 4:1 ratio in frequencies on the bottom line and key phase relations at 45- and 90-degree angles are shown. Lissajous patterns are an example of using a scope's X-Y measurement capabilities.

Other X-Y Measurement Examples

Any time you have two quantities which are not time dependent (unlike phase), you can look at these physical phenomena by use of this X-Y capability. This might include examples of aerodynamic lift and drag or pressure and volumes of both gases and liquids. Remember how we began by stating that transducers take physical phenomena and change them into electric currents or voltages? The preceding examples are applications using this principle.

Component Testers

This is an art in troubleshooting not often used outside on a car. Essentially it is a simpler circuit using your scope's X-Y measurement capability. Refer to Fig. 4-18 and note the 6.3V AC transformer and realize these are out of the circuit component traces. Components tested in circuit have different patterns because of the interconnecting and therefore interaction

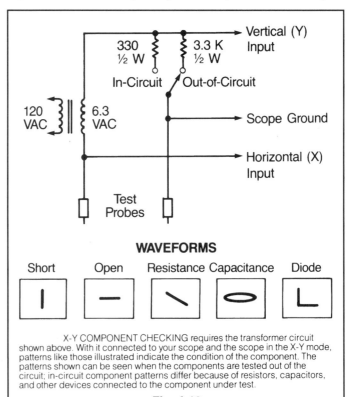

WAVEFORMS

Short   Open   Resistance Capacitance   Diode

X-Y COMPONENT CHECKING requires the transformer circuit shown above. With it connected to your scope and the scope in the X-Y mode, patterns like those illustrated indicate the condition of the component. The patterns shown can be seen when the components are tested out of the circuit; in-circuit component patterns differ because of resistors, capacitors, and other devices connected to the component under test.

Fig. 4-18

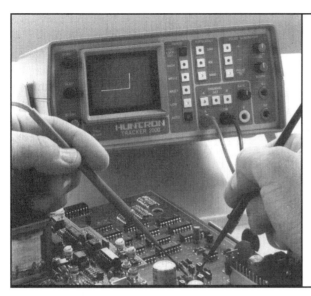

Fig. 4-19
A Huntron Tracker or component tester, in which CRT patterns are crucial to identifying the source of error. (Courtesy Huntron)

of adjacent components, especially resistors and capacitors in series. Figure 4-19 shows a commercial version of this and to be able to discern the pattern "signature" on its CRT takes more practice than I have patience for frankly.

Using the Z-Axis

Remember from our previous discussions that the Z-axis is the intensity axis. That is, higher voltages portray a darker more intense pattern. On some scopes this axis will accept signals up to 100 or more volts and this axis has up to a 5 MHz bandwidth. Positive voltages decrease brightness; conversely, negative voltages increase brightness and a 5 volt change is quite noticeable. The Z-axis is an advantage to you if you have to set up for a long series of tests.

Differential Measurements

Using the ADD and the Channel 2 Invert Vertical Mode allow you to make differential measurements. These type measurements (see Fig. 4-20) allow you to remove unwanted information from a signal

Differential measurements allow you to remove unwanted information from a signal any time you have another signal that closely resembles the unwanted components. For example, the first photos shows a 1kHz square wave contaminated by a 60Hz sine wave. Once the common-mode component (the sine wave) is input to channel 2 and that channel is inverted, the signals can be added by selecting ADD VERTICAL MODE. The result is shown in the second photo.

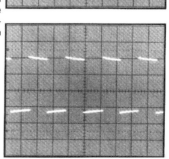

Fig. 4-20

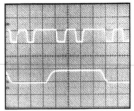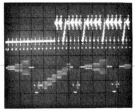

HORIZONTAL ALTERNATE MAGNIFICATION MEASUREMENTS are fast and simple to make. One use, examining timing in a digital circuit, is demonstrated in the left photograph. Suppose you need to check the width of one pulse in a pulse train like the one shown. To make sure you are measuring the correct pulse, you must look at a large portion of the signal. But to measure one pulse accurately, you need a faster sweep speed. Looking at the big picture simultaneously with an enlargement of a small part of the signal is a simple task with the horizontal alternate magnification feature. A second example is shown in the right photograph. Here, triggering is on one field of a composite video signal — displayed by the top (unmagnified) waveform. The bottom (magnified) waveform was attained by setting ALT HORIZONTAL MODE and X50 MAG. With the COARSE and FINE POSITION controls you can walk through the field and look at each line individually.

MEASUREMENTS USING HORIZONTAL ALTERNATE MAGNIFICATION

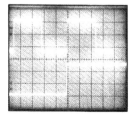

Rise Time

1. Connect your probe to the CH 1 connector and attach the probe tip to the PROBE ADJUST terminal. Hook the ground strap onto the collar of the CH 2 BNC connector and make sure the probe is compensated for channel 1.

2. Use these control settings: VERTICAL MODE to CH 1; CH 1 VOLTS/DIV to 0.2 (10X PROBE); CH 1 variable in CAL detent and pushed in (no vertical magnification); CH 1 input coupling to AC; HORIZONTAL MODE to X1; SEC/DIV to 0.1 ms and its variable control in CAL detent; MAG to X5; TRIGGER SLOPE to the falling edge ( ); MODE to P-P AUTO; SOURCE to either CH 1 or VERT MODE; and COUPLING to AC.

3. If necessary, adjust the TRIGGER LEVEL control for a stable display, then position the waveform in the upper half of the screen. Switch the HORIZONTAL MODE to ALT and use the TRACE SEP control to move the magnified sweep to the lower half of the screen.

4. With the COARSE and FINE POSITION controls, place the rising edge of a pulse on the upper (unmagnified) trace and the rising edge of the corresponding pulse on the lower (X5 magnified) trace along (or nearly along) the center vertical graticule line. Recall that the center vertical graticule line is the registration mark representing the same point in time (from the start of the sweep) along both the unmagnified and the magnified traces. Now your screen should look like the first photo.

Fig. 4-21

---

that closely resembles the unwanted component. As an example, in Fig. 4-20(a), a 1kHz square wave is "contaminated" by a 60Hz sine wave. Once this unwanted signal is input to Channel 2 and that channel is inverted, the signals can be added by selecting the ADD VERTICAL mode. The resulting "clean" signal is shown in Fig. 4-20(b).

Waveform "Expansion"

This can be accomplished by using the SEC/DIV control for displaying the fastest sweep. Remember, the variable control (CAL) knob should be in the fully clockwise position. On some scopes, there is a horizontal magnification mode. These are the ALT and MAG controls, which may be effectively used to enhance the precision on a particular measurement such as rise time (especially) very fast pulses. Refer to Fig. 4-21(a) and note how "innocent" this waveform looks. By using horizontal alternate magnification the "real" waveform appears in Fig. 4-21(b). Further examples are Figs. 4-22(a) and 4-22(b). Note how the rise time is "stretched" out in Fig. 4-22(b) on the lower waveform.

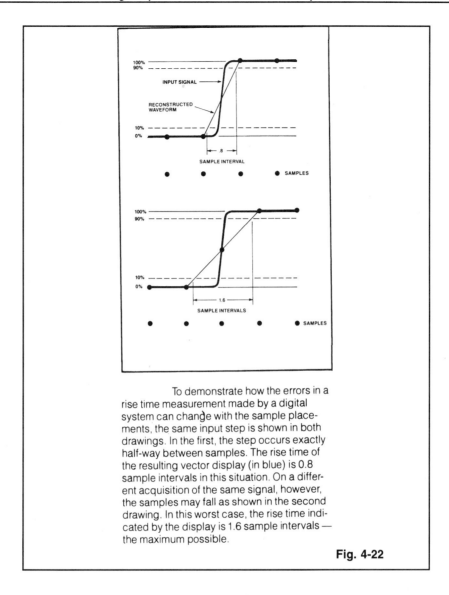

To demonstrate how the errors in a rise time measurement made by a digital system can change with the sample placements, the same input step is shown in both drawings. In the first, the step occurs exactly half-way between samples. The rise time of the resulting vector display (in blue) is 0.8 sample intervals in this situation. On a different acquisition of the same signal, however, the samples may fall as shown in the second drawing. In this worst case, the rise time indicated by the display is 1.6 sample intervals — the maximum possible.

Fig. 4-22

A Closer Look at Rise Time

Previously we took a close look at rise time and defined it. Now, let's try to appreciate what this all important parameter characterizing a square or rectangular wave is.

Square Wave vs. High Frequency Response

Amplifiers are designed with a trade-off in mind between a circuit's high frequency and square wave signal frequency responses. High compensation in a scope's vertical amplifier

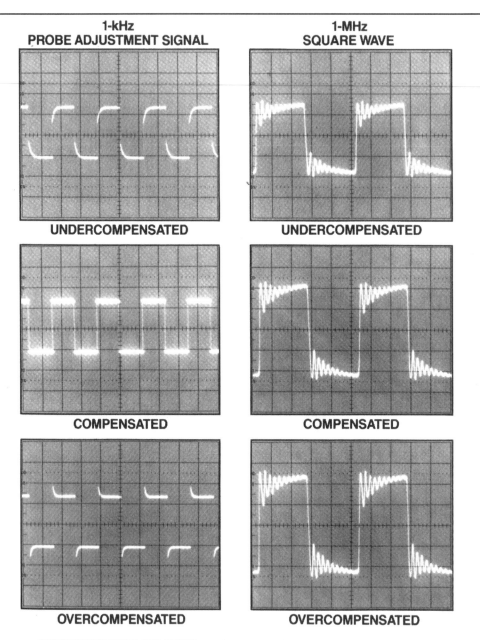

1-kHz
PROBE ADJUSTMENT SIGNAL

UNDERCOMPENSATED

COMPENSATED

OVERCOMPENSATED

1-MHz
SQUARE WAVE

UNDERCOMPENSATED

COMPENSATED

OVERCOMPENSATED

IMPROPERLY COMPENSATED PROBES can distort the waveforms you see on the screen of your scope. In the photographs, the 1-kHz probe-adjustment signal and a 1-MHz square wave are shown as they would appear with proper and improper compensations. Notice the changes in amplitude and ringing on the 1-MHz square wave with differences in compensation

Fig. 4-23

affects the rise time of a square wave measured by the scope. Too much high frequency compensation causes the square waves to overshoot or possibly ring, refer to Fig. 4-23(a). Too little high frequency compensation causes the rise times to roll off the edges of the square wave, refer to Fig. 4-23(b). Proper high frequency compensation, as shown in Fig. 4-23(c), causes a critically damped frequency response.

Bandwidth and Rise Time

The vertical channels of a scope are designed for a broad bandpass, something around DC to the scope's upper limit. This upper limit is defined as the frequency where the amplitude of the input signal has diminished or been reduced to 0.707 of its original value. As an example, a scope with a 35MHz bandwidth will show a 35MHz sine wave with only a -3 dB or 0.707 attenuation. Refer to Fig. 4-24(a) and note that a 15MHz square wave on this 35MHz scope looks pretty bad. This same 15MHz square wave on a 50MHz scope looks much more like a square wave, refer to Fig. 4-24(b).

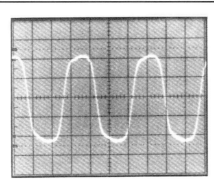

15-MHz
Square wave on a
35 MHz
Oscilloscope

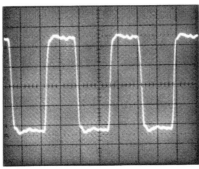

15-MHz
Square wave on a
50 MHz
Oscilloscope

BANDWIDTH SPECIFICATIONS are based on the scope's ability to reproduce sine waves. The upper bandwidth is the frequency at which a sine wave is reduced to 0.707 of the amplitude shown at lower reference frequencies. Though this specification tells you how well the instrument reproduces sine waves, not every signal you examine is sinusoidal. Square waves, for example, have a great deal of high-frequency information in their rising and falling edges that will be lost as you approach the bandwidth limits of the instrument. To illustrate, the two crt photos show a 15-MHz square wave reproduced by a 35-MHz oscilloscope (top) and a 50-MHz oscilloscope (bottom).

Fig. 4-24

The Fluke 97 ScopeMeter®

This DSO (see Fig. 4-25) can do much more than just, for example, a Fluke model 88 DMM. This and the Tektronix (Tek) model 222 are the two DSOs we will use from here on to troubleshoot your car's electrical systems.

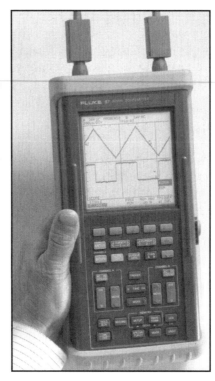

Fig. 4-25
Fluke Model 97
ScoopeMeter®
(Courtesy Fluke
Corp.)

Chapter 4 Quiz

True or False

1. Sensors will soon be placed in automobile transmissions.
2. Scopes analyze electrical signals.
3. A transducer converts electrical energy back into electrical energy.
4. The horizontal beam scans right-to-left.
5. The phosphor coating on the inside of a tradition CRT is what gives it its glowing light producing ability.
6. Your scope's intensity controls adjusts brightness.
7. The beam finder can not override the intensity control.
8. Usually, the trace rotation control is displayed or placed in a prominent position on the front panel of your scope.
9. AC coupling connects the entire AC signal to the input amplifier, and this includes any DC upon which the AC may be riding.
10. A BNC connector is the most common type used on scopes.

Matching questions 11 to 20 with a letter from A to J

11. This control vertical sensitivity.
12. This control increases sensitivity by a factor of 10:1; although, this is at the expense of useful bandwidth.
13. When in this mode, the scope switches back and forth from channel 1 to channel 2 that it tricks your eye into believing that there are no missing gaps in the waveform.
14. To algebraically add channels 1 and 2, you would use these controls simultaneously.
15. If you wanted to see two signals at slower sweep speeds, you'd use this mode of operation.
16. The horizontal system contains a sweep generator that produces a voltage ramp. Its purpose is to control this characteristic of your scope.
17. You can horizontally magnify waveforms by multiplying the sweep speed by the amount of this quantity.
18. This waveform has two amounts of time with two equal states that are symmetrically inverted from each other.
19. This major scope system allows you to precisely define *when* your scope draws its waveforms.
20. These two controls help to precisely define your scope's trigger point.

A. Sine wave
B. Sweep speed
C. CAL
D. Magnification

E. Volts/DIV
F. CHOP
G. Trigger level and slope control

H. CHOP
I. BOTH and ADD
J. Triggering system

Fill in the blanks

21. Your scope's triggering sources are grouped into these two categories _____ and _____.
22. To make amplitude measurements, you fill up most of the CRT _____.
23. _____ measurements are made by performing calculations on the data after you measure it.
24. The term RMS refers to the signal's energy content with respect to producing an equal _____ effect in the same time interval.
25. A _____ pattern helps you determine a signal's phase.
26. Lissajous patterns are examples of your scope using its ___capabilities.
27. The _____ axis is what is responsible for intensity.
28. _____ are devices are designed with a trade off between a circuit's high frequency and square wave signal responses.
29. Too much high frequency compensation causes square waves to _____.
30. The scope's bandwidth is the frequency at which a 1.0V signal would equal this voltage _____.

5

Automotive Sensors:

Diagnosing Them, and Making Vibration
and Rotational Motion Measurements with a DSO

This chapter begins with a discussion of the many sensors found on modern cars. It then progresses into actual procedures used to troubleshoot these sensors. Initially, all troubleshooting is accomplished with a DMM. But later, it concludes with a discussion on how to use a DSO to measure rotational motion and vibration. In between this you will see just how versatile a DSO is in troubleshooting a vast array of areas within your car and its electrical system.

Modern automobiles have one definite characteristic that drastically sets them apart from cars of 20 years ago. This is their vastly improved gas mileage. Even full-sized cars like Lincolns get almost 30 mpg on the highway now. This is only accomplished by optimizing all parameters affecting gas mileage. This includes gas-to-air mixture, temperature, engine rpm, etc. To accomplish this, new cars are laden with a whole host of sensors. Their sole purpose is to feed the engine's computer.

There are three market-driven demands within the automotive sector that can confound a professional mechanic, so you know they will certainly also give you problems. These are:

1. Better performance.
2. Better fuel economy.
3. The absolute compliance with state and federal regulations concerning emissions.

Couple this with increasingly sophisticated electronic components (especially sensors) found on modern cars to achieve these laudable goals and you have a good understanding of the environment in which a mechanic has to operate today.

Government regulations, competing international designs, and higher customer expectations have caused cars to be tweaked to where their computer-controlled engines are more efficient and adaptable. Today's multiport fuel injection system precisely dispenses gas vapor and computer ignition systems time the spark. What is missing is an accurate determination of the mass of the air within the hostile environment of the intake manifold. The ideal sensor would allow the engine's computer to instantly adjust the air and fuel concentrations to the optimum ratio of 14.7:1 under all driving conditions!

MAP Sensors

This air or oxygen sensor has to be small, tough, and retain its accuracy in such diverse conditions as cold starts, high altitudes, changing loads, and quick accelerations. Presently, carmakers and sensor companies are spending millions of dollars in researching this superior air detector. Today, airflow is mathematically inferred by measuring key variables in the manifold. After sensing Manifold Air Pressure (MAP), air temperature, engine rpm, and the mass of the air entering the cylinder, a calculation is made by the car's computer. However, accuracy suffers, especially under varying conditions, due to an inherent time lag between the computation of these parameters and their arrival to your car's computer.

There are mechanical sensors that directly gage airflow; however, they are large, expensive, and unreliable. In such an arrangement (see Fig. 5-1), a half-inch frame holds two heat elements in the airflow path. Air mass is measured by the amount of heat it takes to maintain one element's temperature. The elements consist of platinum wire wound around a quartz bobbin or ceramic bobbin.

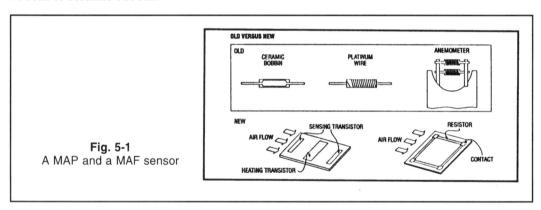

Fig. 5-1
A MAP and a MAF sensor

As air passes the element, it evacuates heat in much the same manner as the wind-chill effect upon the moist lips of a downhill skier. The difference in temperature is a direct analogy between the current in the two elements.

These "anemometers" are already in use is several European luxury and sports cars. Bosch and Siemens use them in their fuel injection systems. But the cost of $30 severely limits their use in mass-produced cars.

Other companies such as Hitachi, Ford, Honeywell's Microswitch division and Teknetron are all working toward solving this problem. Their goal is not to produce a MAP but rather a mass air-flow (MAF) sensor built of silicon, again see Fig. 5-1.

Motorola already sells more than 20 million MAP sensors and MAF sensor makers hope to sell at least that many. The market will be driven by such factors as California's on-board diagnostic regulations in cars. This went into effect in 1996. This regulation also pushes for more sensors to be placed in transmissions and the chassis.

The newer silicon technology meeting this challenge consists of a single piezoceramic element. It is positioned to maximize the sensitivity to shear stress at an angle of 45 degrees near the edge of the 25-micron-thick (a micron is a millionth of an inch) diaphragm etched into the silicon.

This sensor provides an output in the millivolt range in direct proportion to applied pressure. More recent advances have also allowed the temperature compensation and amplification sections to be placed right on the same chip.

A Real-World Troubleshooting Example

The following example illustrates the process used to diagnose a problem related to a computer engine control system. It doesn't cover all possible causes, just the most probable. Ford calls the computer the ECA, and the O_2 sensor the EGO sensor. We refer to them is this chapter by the Ford designations and also as the ECM and O_2 sensor.

A 1985 2.3L (liter) carbureted Mustang has the EEC-IV control system. The complaint is that the fuel economy has been poor; it doesn't have much pep.

Since this type of problem could have causes not related to the computer you first put the car on the ignition scope and the exhaust gas analyzer for as overall system check.

Fig. 5-2
Fluke models 86 and 88 autmotive DMMs

A look at the scope tells you that there are two fouled spark plugs and the exhaust gas analyzer tells you that the CO, at 14 percent, is more than 10 times too high. The CO level tells you that you have found the cause of the customer's complaint—a rich fuel mixture. But what's making it run so rich?

Start by initiating the diagnostic procedures. By using a digital multimeter (DMM) with an analog bar graph (see Fig. 5-2), setting the meter to DC volts, and watching the bar graph movement, you can access the code displays for the EEC-IV system. You discover a code 42, which indicates a rich condition, but you already knew it was too rich. So how do you detect and read an error code like the 42 we read for our example? Note in Fig. 5-3 that when a sensor fails, a code is stored in memory. You can read this code by triggering the memory. Carmakers provide various ways to read a fault code;

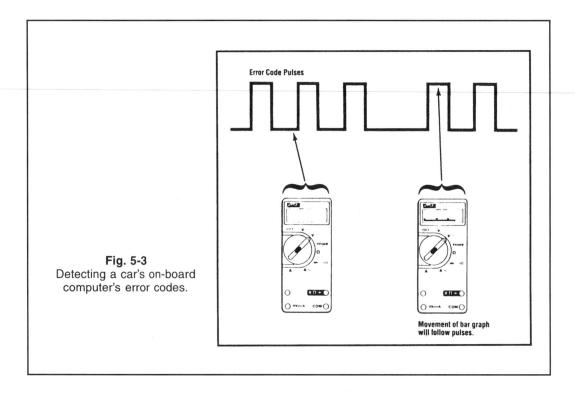

Fig. 5-3
Detecting a car's on-board
computer's error codes.

sometimes the digital clock displays it, others use the tachometer or a blinking light. From Fig. 5-3 you can read the code by looking at the pulses at the bottom of the DMM's display (note the DMM on the right in Fig. 5-3). You must use a DMM with at least a 10 Megohm input impedance to avoid blowing the delicate circuitry associated with the car's computer. A mere resistance check with an analog meter mistakenly applied to the car's computer will damage it. Ford specifies an analog meter to read these pulses from memory, but put it away after reading these pulses or use a DMM like the Fluke DMMs used within these examples because they have a bar graph at the bottom of their scales.

There are many things that may set a code 42. While cleaning the spark plugs, you make a mental list in order of difficulty to check, least difficult first:

1. Low coolant level
2. Engine oil contaminated with gasoline
3. Dirty air filter
4. Low vacuum—worn engine, vacuum hoses cracked, especially the BP/MAP hose
5. Carbon canister saturated with gas
6. Corroded coolant-sensor harness connection
7. Corroded air-charge temperature sensor harness connector
8. Open EGO sensor signal wire
9. Contaminated EGO sensor
10. Carburetor faults—sunk float, leaking needle valve, etc.

There are other problems that would cause a rich mixture, but they would set a combination of codes.

You start checking the items on the list, making sure the coolant sensor is properly submerged in coolant. You find the coolant level is OK. Next you pull the dipstick and sniff the oil checking for gasoline contamination. It smells a little like gas so you decide to check further.

You disconnect the EGO sensor from the harness and connect a Fluke 88 DMM between signal and engine ground. Set the meter to volts DC, 4V range and start the engine. Now, by removing the PCV valve while you watch the EGO voltage on the meter you can determine if the engine is getting excess gas fumes from the oil. If the EGO voltage starts out high and drops when you pull the PCV valve you can conclude that the crankcase was the source of the extra fuel vapors. The voltage stays high. That wasn't the problem.

Next, pull the air filter to check for dirt; OK here, too.

Inspect the vacuum hoses for cracks and leaks that could fool the BP/MAP sensor, and check the hose to the sensor, making sure it's connected. No problems here, either.

Since the DMM is still hooked up to the EGO you pinch the canister purge hose while watching the EGO voltage, checking for a saturated carbon canister as a source of fumes. Again, no change. The canister's not the problem, either.

You haven't checked the EGO wiring or the temperature sensors yet. You can kill two birds with one stone by installing your Breakout Box (BOB) between the ECA and the wiring harness now.

Caution: Always use the BOB rather that piercing the wire or probing the connector directly because you may damage the delicate connectors in the plug causing more problems than you solve.

Since you've left the EGO sensor unhooked, now is the time to check the wiring for opens in the EGO circuit. You install a jumper on the EGO harness connector between the signal and signal return pins, looking for a short between the appropriate pins on the BOB. You get a reading of 2 ohms, showing there's not a problem. You remove the jumper and reconnect the EGO sensor.

Check the resistance of the coolant temperature and air-charge sensors. You also check the temperature at the base of each sensor and compare the resistance values for each sensor to the values shown for those temperatures found is the manufacturer's specification tables.

Both sensors have more resistance than they should. But is the problem the sensor or the wiring? You unplug the connector at each sensor to measure the resistance across the sensor terminals and notice green corrosion on both sensors. To be sure this is the problem you clean the terminals, reconnect the sensors, and check the resistance of each sensor at the BOB. The readings are normal.

You remove the BOB, reconnect the ECA, disconnect the EGO sensor, ground the signal wire with a jumper and run the engine at a fast idle for a couple of minutes to clean the EGO sensor. Next, you run the self-test. The car passes with no codes and on a test drive it runs fine.

Danger

In modern electronically controlled ignition systems there are dangerous voltages present. Is some cases there is enough energy to kill. Observe the following precautions:

1. Be sure to switch off the ignition system when doing the following:
   - When replacing spark plugs, the ignition coil, or high-tension leads.
   - When connecting ignition-related test equipment, timing lights, scopes, dwell/tachometers, etc.
2. Do not remove spark plug wires or leave them disconnected with the ignition switch on. It could damage components such as the ignition coil or module. The high voltage can cause serious injury.
3. When checking for the presence of spark, use an ignition voltage tester—don't just pull off the spark plug wire.

Oxygen Sensors

The O_2 sensor sends its voltage information to the ECM, which controls injector on-time or carburetor duty cycle (dwell) to maintain the optimum mixture.

Zirconia sensors act as batteries, generating a voltage output dependent on the amount of oxygen it is sensing. Titania O_2 sensors are used in some applications where its output is in the form of resistance that changes its response to oxygen level changes.

Near the Stoichiometric (ideal) air/fuel ratio (14.7:1) the O_2 sensor responds to small mixture changes with large voltage changes (see Fig. 5-4). The sensor will not tell you the amount of oxygen in the exhaust, only whether or not there is oxygen. If there is no oxygen present (rich condition) the voltage will rise to about 800 mV. When there is a large amount of oxygen present (lean condition) the voltage will fall to about 200 mV.

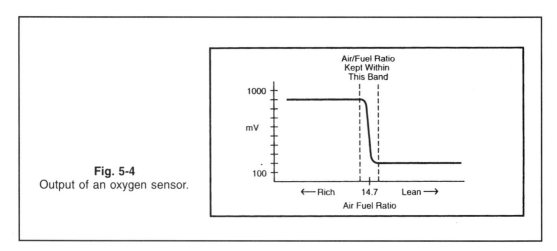

Fig. 5-4
Output of an oxygen sensor.

If the time during which there is oxygen present is the same as the time when there is no oxygen (50 percent) then the voltage at the oxygen sensor will *average* 500 millivolts (mV). If the time during which there is no oxygen present (rich condition) is greater than 50 percent, then the *average* voltage will be greater than 500 mV, indicating a generally rich condition. Likewise, if the time during which there is no oxygen present is less than 50 percent, then the *average* voltage will be less than 500 mV, indicating a generally lean condition. By design, the system oscillates back and forth between "oxygen" and "no oxygen," so the oxygen sensor voltage oscillates roughly between 800 mV and 200 mV.

When sensors get dirty and need replacement, the high and low voltages get closer together. Some service experts recommend replacing them if the voltage range fails to go below 300 mV and above 600 mV. Remember, the system must be in closed-loop (600°F exhaust temperature) before you'll see any oscillation.

Another method of checking for a "lazy O_2 sensor" is by watching cross counts. The oxygen sensor should produce readings that alternate between rich and lean every couple of seconds. Replace the sensor if readings do not change back and forth quickly or if the sensor shows a fixed lean condition. An analog bar graph or pointer is extremely useful in watching for these quick voltage changes.

Manifold Absolute Pressure Sensor (MAP)

To accurately control fuel mixture and ignition timing the ECM must monitor the air mass entering the cylinders. Air mass is affected by volume and density. Greater volume or density means greater mass. One method of finding air mass is to measure it directly with a Mass Airflow sensor. Another way is to measure it indirectly and calculate mass from the following information:

1. Engine rpm (determines air volume)
2. Temperature (affects air density)
3. Manifold Absolute Pressure (MAP, the only other factor to determine air density)

When the manifold absolute pressure is low, engine vacuum is considered to be high. At wide-open throttle the MAP (Manifold Absolute Pressure) is high, and at low rpm with a small throttle opening (such as when coasting downhill), MAP is low.

The ECM knows rpm from the engine speed sensor (crank position sensor often supplies this information) and Manifold Absolute Pressure from the MAP sensor. From this information it determines the air volume entering the cylinders and the air density due to altitude and barometric pressure. However, temperature also affects air density without affecting MAP. So, an Air Charge Temperature sensor (ACT) gives the ECM this final piece of the puzzle.

Why not just measure vacuum?

Manifold vacuum is the difference between the air pressure inside the manifold and outside. Because the ambient air pressure is lower at high altitude, a car climbing a 6 percent grade at 50 mph would have a lower manifold vacuum reading at 5,000 feet than the same car climbing a 6 percent grade at 50 mph at sea level. In both cases the engine would be producing the same power and using the same mass of air.

If the ACT is the same, the MAP would be the same is both cases. If the ECM was measuring manifold vacuum only, it would think the engine was under heavier load because vacuum was lower. Normally under high load, the ECM commands richer mixture and retarded spark, but in this case it would retard the spark too much and enrich the mixture too much. By measuring the absolute pressure inside the manifold the ECM can tell what the load is regardless of altitude or barometric pressure. Barometric pressure also affects manifold absolute pressure in exactly the same way as altitude.

The MAP sensor will give the ECM correct information about engine load and barometric pressure regardless of altitude. So the correct full name of this type of device is just as Ford describes it—Barometric Pressure/Manifold Absolute Pressure sensor (BP/MAP sensor).

Ford BP/MAP sensors differ from other MAP sensors, in that they output a 5V square wave rather than an analog voltage. The frequency of the square wave changes depending on manifold vacuum, altitude, and changes in barometric pressure. When the engine is working hard at wide-open throttle the MAP is high. At low rpm and wide-open throttle, the MAP will be nearly the same as outside air pressure (relatively high). On the other hand, at small throttle opening (such as when coasting down hill) the MAP is relatively low.

Mass Airflow Sensors (AC-Delco Type) (MAF)

The AC-Delco MAF sensor shown in Fig. 5-5 is a form of "hot-wire anemometer" that measures air volume and density, as we have already seen. It produces a frequency output. Other manufacturers also use MAF sensors, some of which have a frequency output and some of which have an analog voltage output. They all work basically the same way.

A constant voltage source is applied to the hot wire. The wire has a positive temperature

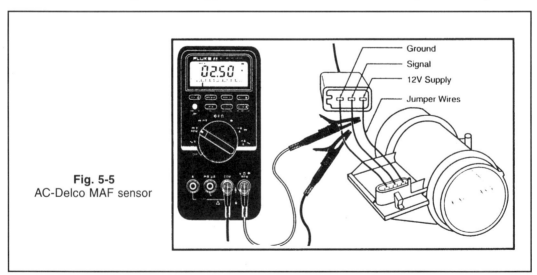

Fig. 5-5
AC-Delco MAF sensor

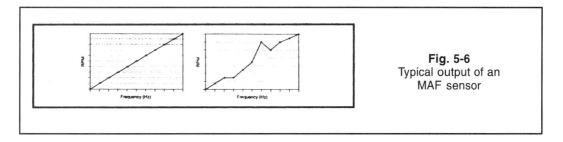

Fig. 5-6
Typical output of an
MAF sensor

coefficient, meaning, that as it gets hotter its resistance increases. The incoming air tends to cool the wire, lowering its resistance, thereby increasing the current. Hot dry air, being less dense (and having less mass), cools it less than cool moist air, which is more dense (and has more mass).

The greater the air mass passing the hot wire, the greater the current flow. A circuit mounted on top of the AC-Delco MAF sensor converts the current flow into a square wave whose frequency changes depending on the mass of the airflow. Figure 5-6 is a typical output for a working MAF sensor.

Cam and Crankshaft Position Sensors

Crankshaft and camshaft sensors tell the ECM the position of the crankshaft and camshaft, and when TDC #1 occurs. From this information the ECM knows when to fire spark plugs, when to pulse injectors, and even provides an rpm signal. These sensors are close relatives of those found in distributors and function in the same way. There are two types, Hall-effect and magnetic.

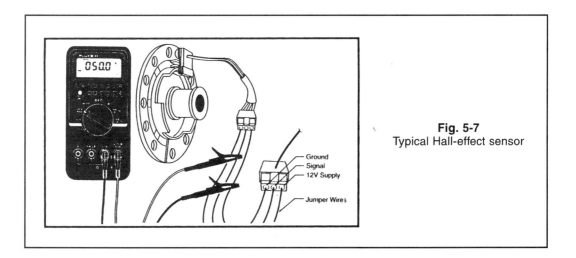

Fig. 5-7
Typical Hall-effect sensor

Hall Sensors

The Hall sensor (see Fig. 5-7) is found on late-model GM cars. Hall sensors typically output a 5V square wave pulse each time the magnetic field from the permanent magnet is interrupted by the vane wheel.

Crisp leading and trailing square wave edges are very important to precisely tell the computer when to respond. If the Hall transistor gets weak or out of adjustment, the edges are not distinct and timing becomes erratic.

Magnetic Sensors

Magnetic or variable reluctance sensors work much like a magneto, except that on a position sensor, the magnetic field fluctuates instead of wire coils moving as in a magneto. The output is an AC pulse, whose voltage rises in proportion to engine speed. Of course, the frequency also rises, telling the ECM engine speed as well as cam or crankshaft position.

The pulse voltage can vary from 0.1V while cranking to 0.5V at idle to 5.0V or more at road speed. Magnetic type ABS wheel speed sensors also work this way.

Fuel Injection On-Time

Electronic fuel injectors are controlled by the ECM and influenced by a variety of operating conditions, including temperature, engine load, and feedback from the O_2 sensor during closed-loop operation.

Fuel injection on-time can be expressed as a percent of total time (duty cycle) or in milliseconds (pulse width) and indicates the amount of fuel delivered to the cylinder. Greater pulse width means more fuel, provided fuel pressure stays the same.

Two different driver transistor circuits are used to control the ground path of the injector: saturated switch (conventional port fuel injection systems) and current regulated (peak and hold).

In conventional port fuel systems, the ECM provides a ground path for the injector through a driver transistor. When the transistor is "on," current flows through the injector winding and the transistor to ground, opening the injector valve.

With current regulated systems, the current through each driver transistor is allowed to rise to a "peak" of 4 amps quickly, unseating the injector pintle starting fuel flow, at which time the injector current is reduced to 1 amp to "hold" them open. Some manufacturers refer to this as a "peak-and-hold" injector system, which has a particular advantage in the upper speed range.

The Fluke 88 DMM is designed to accurately display the millisecond on-time of a typical port fuel injector depicted in Fig. 5-8 Diagram A; Fig. 5-8 Diagram B depicts a current-regulated fuel injector and may require the use of an oscilloscope to correctly measure milliseconds on-time due to the multiple voltage spikes present.

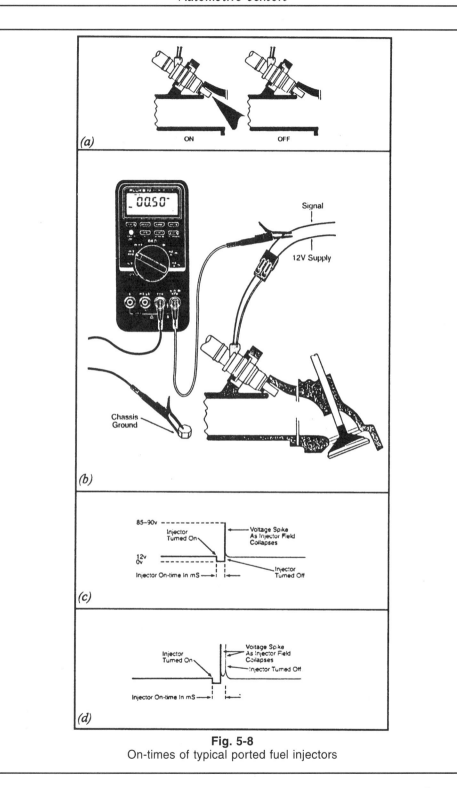

Fig. 5-8
On-times of typical ported fuel injectors

The amount of fuel delivered to a cylinder can be affected by reduction in current flow. Low battery voltage, resistance in connectors, a weak driver transistor, or resistance in the computer's ground will also tend to cause a lean condition.

The voltage at the ECM return side of the injector will fall to nearly zero when the injector is on. When the injector is off the voltage will be battery voltage. For this reason we are interested is negative pulse width.

When the engine is turning at high speed and under heavy load, the injector may be on for nearly a full crankshaft revolution. Under extreme conditions it may barely have time to close before it must open again. If the current flow through the injector is too low (low battery voltage, for example) the injector may open too slowly, causing a lean condition (see Fig. 5-9).

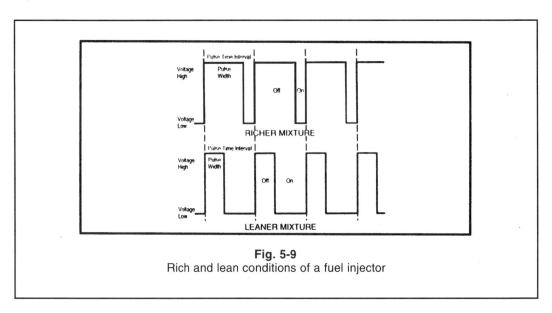

Fig. 5-9
Rich and lean conditions of a fuel injector

Feedback Carburetor

Feedback Carburetors are controlled by the electronic control module in response to feedback voltage from the O_2 sensor. The ECM sends out a square-wave signal with a constant 10 Hz frequency and voltage, but a variable ratio of off-time to on-time (duty cycle) once in closed-loop operation. The duty cycle depends on the O_2 sensor voltage.

The square-wave signal typically controls a solenoid that, in turn, meters air and fuel to adjust the fuel mixture within limits set by jet and air-bleed size. Typically, the ECM supplies the electrical ground for the solenoid through a driver transistor. When the transistor is on, current flows through the solenoid causing it to move to a lean position.

The duty cycle (or six-cylinder dwell equivalent) can be measured and can indicate proper system operation. When measuring duty cycle we are interested in negative duty cycle (the time the signal voltage is low) because this is when the driver transistor is on, providing a current path to ground for the solenoid. During this time the solenoid will be on. Usually, *on* means the metering rods are in the jets obstructing fuel flow or the air bleeds are open (or both on some Rochesters) causing a lean condition. For these situations, greater negative duty cycle means the ECM is issuing a lean command. Forcing a lean or rich condition should cause duty cycle to change if everything is working correctly.

The fuel metering rod is a typical Rochester-type carburetor, like Motorcraft and Holly types, has a control solenoid that gets a square-wave signal from the ECM. The Rochester carburetor shown in Fig. 5-10 is a solenoid-controlled feedback type used in many GM applications.

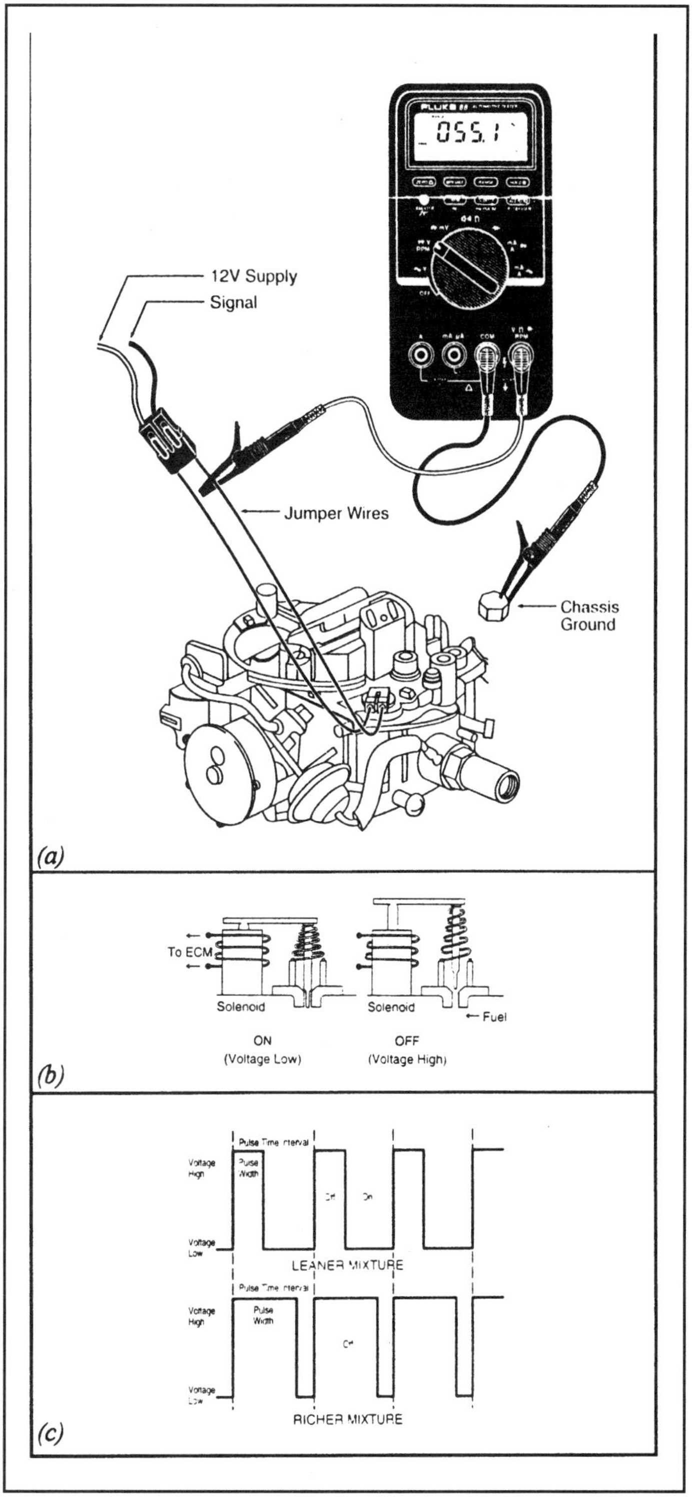

Fig. 5-10
Solenoid-type feedback
Rochester carbuertor

Idle Air Control (IAC) Motors

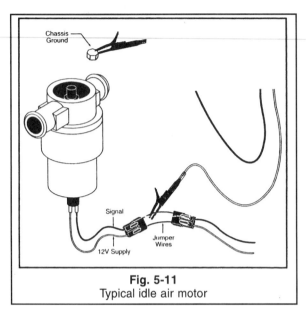

Fig. 5-11
Typical idle air motor

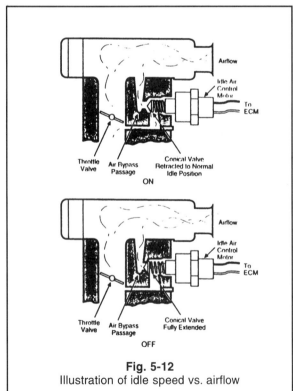

Fig. 5-12
Illustration of idle speed vs. airflow

Sometimes called Idle Stabilization or Idle Speed Controls (ISC), there are two electronic types. One relies on current control to hold a position, and the other is a stepper motor that moves a given distance for each signal it receives from the ECM. Except for basic resistance tests, stepper motors must be monitored with a scan tool.

A typical idle air control motor is shown in Fig. 5-11. A control valve, much like an oversized fuel injector valve, moves back and forth to control airflow. More air tends to increase the idle speed compensating for additional loads, see Fig. 5-12.

The IAC motor (to be discussed) is a Bosch rotary idle actuator found on late-model BMWs. It rotates only 60 degrees between fully open and fully closed. As it rotates it uncovers an air bleed port allowing air to bypass the throttle plate.

Movement is controlled by pulsing DC current through the motor. The ECM pulses the current on and off at a fixed frequency but varies the ratio of "on-time" to "off-time" (duty-cycle). The more time the current is pulsed on, the higher the average current flow. The idle motor responds to the average current, moving a distance proportional to the amount of current. More current equals higher idle speed.

Current control types can be monitored by measuring the duty cycle of the signal the ECM sends to the IAC motor, or by measuring the average DC voltage, or by measuring the average current.

Temperature Sensors

Virtually all temperature sensors used in computerized engine controls are thermistors, having a negative temperature coefficient (NTC). This means that, as the sensor's temperature increases, its resistance decreases.

If you graph the temperature vs. resistance, you will see a curved line that is very steep at low temperature and flattens out at higher temperatures (see Fig. 5-13). Table 5-1 shows

Table 5-1
A negative temperature coefficient (NTC) thermistor's resistance vs. temperature

NTC Thermistors	
GM (All)	
°F	Resistance in Ω
210	185
160	450
100	1,600
70	3,400
40	7,500
20	13,500
0	25,000
-40	100,700
Ford Air Charge & Coolant Temperature Sensors	
°F	Resistance in Ω
50	58,750
65	40,500
180	3,600
220	1,840
Chrysler 3.0 Liter Coolant & Air Charge Sensors	
°F	Resistance in Ω
70	7,000 to 13,000
200	400 to 1,500

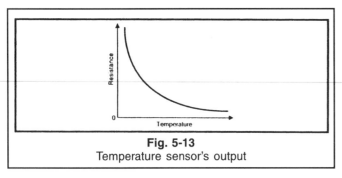

Fig. 5-13
Temperature sensor's output

some NTC thermistors used by American manufacturers. Temperature sensors that have drifted out of range or that have corroded connections can cause rich or lean conditions. An open-circuit coolant temperature sensor can cause the vehicle to stay in open loop.

Throttle Position Sensors

A throttle position sensor is simply a variable resistor connected to the throttle shaft. Some people think of it as a replacement for an accelerator pump on the throttle body or port fuel injected engines. But it is much more than that. Its function is to tell the ECM how far the throttle is open, whether it is opening or closing, and how fast it is opening or closing. As its resistance changes, so does the voltage signal returning to the ECM. It can be tested either by watching the voltage change or by watching the resistance change, using the analog pointer on the Fluke 88. Some throttle position sensors are digital. At idle, one set of contacts is closed, and at wide open throttle, another set of contacts is closed. When testing sensors, be sure to check the manufacturer's specifications for the proper application. The internal workings of typical throttle position sensors are shown in Fig. 5-14.

Now let's look at engine components further, predominately sensors.

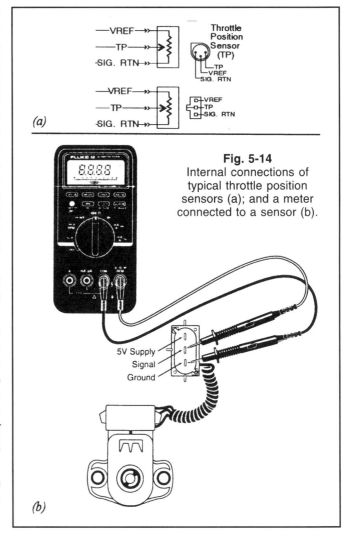

Fig. 5-14
Internal connections of typical throttle position sensors (a); and a meter connected to a sensor (b).

Sensors

Oxygen Sensors

Note: The engine must be running at operating temperature (600°F exhaust temperature) and in closed-loop mode before starting this test.

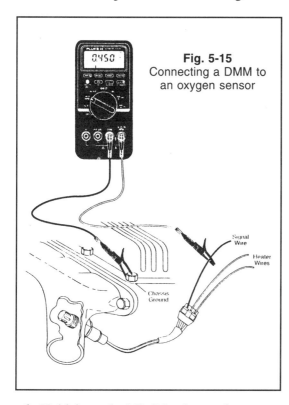

Fig. 5-15
Connecting a DMM to
an oxygen sensor

Min/Max Average Test for O_2 Sensors, Key-On-Engine-Running

1. Connect the DMM to the O_2 sensor wire as shown in Fig. 5-15. Be sure you are connected to the signal wire and not to one of the heater or ground wires. Set the meter to read DC volts by turning the rotary knob to VDC and set the range to 4 volts by pushing the range button three times.

2. Watch the voltage oscillate to be sure you are in the closed-loop mode (100 mV to 900 mV).

3. Press the MIN/MAX button once to enter the DMM's min/max/average function.

4. Let the engine run for a few minutes to give the DMM a chance to record enough readings.

5. Press the Min/Max button slowly three more times while watching the display. You should see a Maximum of more that 800 mV, a Minimum of less that 200 mV, and an Average of around 450 mV.

6. Hold down the Min/Max button for two seconds, then push it once more to reset the Min/Max function.

7. Drive the system rich using propane enrichment.

8. Repeat steps 4 and 5 to read the average voltage. Average O_2 voltage should be higher, indicating a rich condition.

9. Repeat step 6 to reset Min/Max.

10. Disconnect a large vacuum hose to create a lean condition. Partially block the vacuum leak if the engine tends to stall.

11. Repeat steps 4 and 5 to read average voltage. Average voltage should be lower, indicating a lean condition.

Note: Exhaust temperature must be greater than 600°F except for heated O_2 sensors. Many fuel-injected engines will react too fast for this test to be effective; the engine rpm will just rise and fall.

On-Road Tests

By driving the car with the DMM hooked up to the O_2 sensor you can identify faults in other systems.

Note: Engine must be at operating temperature.

1. Temporarily fasten the DMM to a stable area under the hood. Connect the meter to the O_2 sensor.
2. Set the meter to read DC volts by turning the rotary knob to VDC and set the range to 4 volts by pushing the range button three times.
3. Drive the car. Do not turn the engine off or disconnect the DMM when returning.
4. With the engine running, press the MIN/MAX button three times to see the maximum, minimum, and average readings that were recorded.
5. High average voltage means too much fuel. For example: ECM is commanding full lean but cold start valve is leaking.
6. Low average indicates too little fuel. For example: ECM is commanding full rich, but can't compensate for vacuum leak.

Ford BP/MAP Sensor

DC Voltage Test For Intermittence

Key-On-Engine-Off

1. Set up the DMM by turning the rotary dial to V/RPM.
2. Measure the DC voltage. It should be 2.5V. Tap sensor with small screwdriver while watching for intermittence. Repeat tapping to induce intermittence, while heating the sensor with a hair dryer.

Frequency Measurement

Key-On-Engine-Off

3. Push the rpm button three times so the meter displays Hz.
4. Measure the frequency. It should correspond to the chart on the right depending on your altitude.
5. Using a hand vacuum pump, draw a vacuum while watching the frequency. Graph your results. Compare with those is Fig. 5-16.

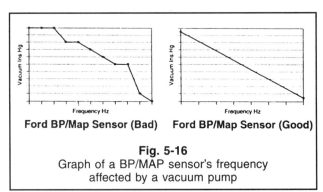

Ford BP/Map Sensor (Bad)   **Ford BP/Map Sensor (Good)**

Fig. 5-16
Graph of a BP/MAP sensor's frequency affected by a vacuum pump

AC-Delco MAF Sensor

Note: GM, Ford and many foreign manufacturers use Bosch-type sensors that output as analog voltage. Make sure you are working on a Delco unit for this test.

DC Voltage Test for Intermittence

Key-On-Engine-Running

1. Set up meter to measure DC volts by turning the rotary dial to V/RPM.
2. Start the engine.
3. You should see a steady 2.5V.
4. Tap on the sensor with a small screwdriver handle.
5. You should see no fluctuation of the analog pointer, nor should there be any engine misfire when you tap.
6. Repeat the test, tapping for intermittence while heating MAF sensor with a hair dryer.

Frequency Measurement Over a Range of rpm

Key-On-Engine-Running

1. Set the DMM to DC volts.
2. Press the rpm button three times so the meter displays Hz.
3. Start the engine.
4. You should see a frequency of around 30 Hz with the engine at idle (depending on which engine you're working on).
5. Measure the frequency at various engine speeds and record your results.
6. Make a graph (Fig. 5-17).

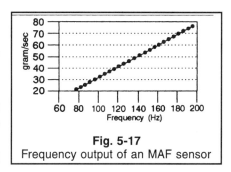

Fig. 5-17
Frequency output of an MAF sensor

Cam And Crankshaft Position Sensors, Hall Sensors

Test For Presence of the Square-Wave Signal

Key-On-Engine-Off

1. Set the meter to measure DC volts (see Fig. 5-18) by turning the rotary knob to V/RPM.
2. Disconnect the ignition primary to disable the engine ignition. While cranking the engine watch for pulses of the analog pointer each time the vanes pass the sensor. Note: At idle or faster the pulses are too fast for the bar graph.
3. Press the Duty Cycle button once. Duty Cycle can be an indication of square-wave quality. Poor quality sig-

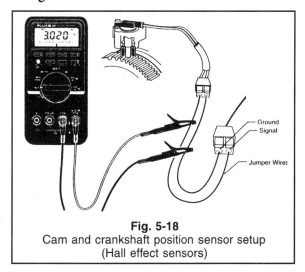

Fig. 5-18
Cam and crankshaft position sensor setup
(Hall effect sensors)

nals will have a low duty cycle. Good Hall sensors should have a duty cycle of 50 percent, depending on the type of sensor. Measure a few known good cars and record the results for future reference.

Fuel Injection On-Time

Pulse Width Measurement of Fuel Injector On-Time

Key-On-Engine-Running

1. Set up the meter to measure DC volts, as shown in Fig. 5-19, by turning the rotary knob to V/RPM.
2. Press the Duty Cycle button twice to get into the Pulse Width function.
3. Read the Pulse Width on the DMM display. (If pulse width seems low and not increasing in response to throttle increase, this may indicate a current-limiting injector system).

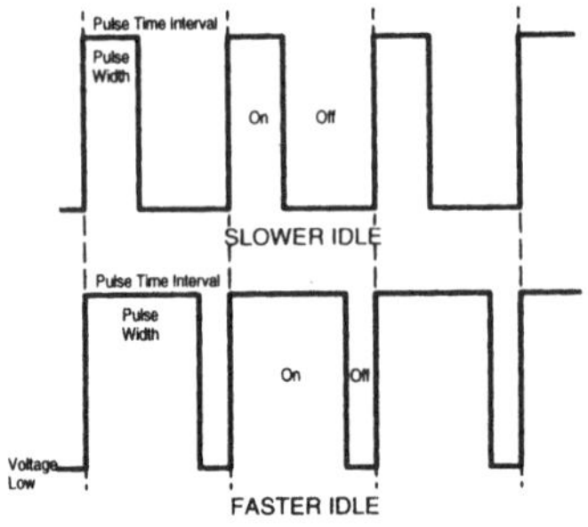

Drive the System Rich

1. Disconnect the O_2 sensor.
2. Grasp the connector coming from the vehicle harness with your fingers (you must make electrical contact).
3. With your other hand, touch a good clean ground (this is a safe procedure for low voltage, low current circuits).
4. Read the injector pulse width from the DMM display.
5. The ECM will interpret this as a lean condition (low O_2 voltage) and compensate with a rich command. Pulse width will increase.

Drive The System Lean

1. Now remove your finger from ground and touch a battery voltage point (battery post or some accessory source).
2. Read the Pulse Width on the DMM display.
3. The ECM will interpret this as a rich condition and compensate with a lean command. Pulse width will decrease.

Fig. 5-19
Fuel injection
on-time/pulse width setup

Cam and Crankshaft Position Sensors
Magnetic Sensors

Test For Presence of Sine Wave Signal

Key-On-Engine-Off

1. Set up the meter to measure AC voltage by turning the rotary knob to ACV.
2. Disconnect the ignition primary to disable the engine ignition. Crank the motor while watching the display. You should see about 0.1 volts AC on systems with many "teeth" on the pickup wheel. On systems with only a few "teeth" you will need to switch the meter to the DC volts function and watch the analog pointer jump whenever a "tooth" comes by.

Feedback Carburetors

Note: The engine must be at operating temperature and in closed-loop mode.

Duty Cycle Test For Feedback Carburetors

Key-On-Engine-Running

1. Connect the meter to the green Mixture Control (MC) test connector at the base of the carburetor. (See manufacturer's component locator.)
2. Set up the meter to DC volts by turning the rotary knob to V/RPM. Push the duty cycle button once.
3. Start the engine.
4. Observe the duty cycle reading. It should fluctuate within a range of about 20 digits (0.20%) if the system is in closed-loop mode. A reading of 50 percent would be normal in most applications.
5. In some applications if you see OL on the display connect the red test lead to the other wire.

Drive the System Rich

1. Drive the system rich using propane enrichment, but be careful not to stall the motor by adding too much.
2. The duty cycle should rise as the electronic control module commands full lean to compensate for the rich mixture.

Drive the System Lean

1. Pinch the PCV hose with your fingers and pull it off the PCV valve.
2. As you release the pinch on the hose the duty cycle will drop as the electronic control module tries to compensate for the vacuum leak. Don't let in too much air or you'll stall the motor.

Note: The above technique also works on port fuel injected engines and throttle body injected engines.

Idle Air Control (IAC) Motors

There are three different tests for idle air control motors:

- Measure the average DC voltage.
- The duty cycle (which controls the average voltage and current).
- The current, depending on what the manufacturer specifies.

Average Voltage Test

Key-On-Engine-Running

1. Connect the meter to read voltage between the IAC motor's return wire and ground as shown. Set the meter to measure DC volts by turning the rotary knob to V/RPM.
2. Start the engine and observe the reading on the DMM display.
3. A normal idle speed with lower than manufacturer-specified voltage can indicate a vacuum leak. A higher-than-normal voltage can indicate high accessory load, engine mechanical problems, or a sticking IAC motor.

Duty Cycle Test

Key-On-Engine-Running

1. Connect the meter to read voltage between the IAC motor's return wire and ground.
2. Set the rotary dial to DC volts.
3. Press the Duty-Cycle button once to set the meter to read duty cycle.
4. Start the engine and observe the reading on the display.
5. A normal idle speed with lower-than-specified duty cycle can indicate a vacuum leak. A higher-than-specified duty cycle can indicate high accessory load, engine mechanical problems, or a sticking IAC motor.

Average Current Test

Key-On-Engine-Running

1. Connect the meter in series with either wire leading to the IAC motor. Install the test leads in the 10 Amp and common jacks.
2. Set the meter to measure DC current by turning the rotary knob to DC Amps.
3. Start the engine.
4. The meter will read the average DC current.
5. As with the duty cycle-method, high average current can indicate high accessory load or mechanical problems; low average current can indicate vacuum leaks.

Warning: Always return the test leads to the V and COM jacks after making current measurements to avoid damage to the meter and the components under test.

Temperature Sensors

Note: Remember to check coolant level in the vehicle radiator when cold before you start this test. Coolant sensors must be immersed in water to work properly.

Temperature Test for GM, Chrysler, and Other Systems Providing Serial Data Output

Key-On-Engine-Running

With Scan Tool: Compare Scan Tool data with temperature measured at sensor using a Fluke 80TK with an 80PK-6A probe (see Fig. 5-20).

1. Set up your scan tool to read the value of the temperature sensor.
2. Set up the DMM and temperature module by turning the rotary knob to mV DC and plugging the module into the meter.
3. Switch the module "on" to °F or °C as needed.
4. Touch the tip of the temperature probe to the base of the sensor.
5. The temperature you measure should be within about 10°F of the temperature you read from the scan tool. If not, the sensor has drifted out of spec, there is a

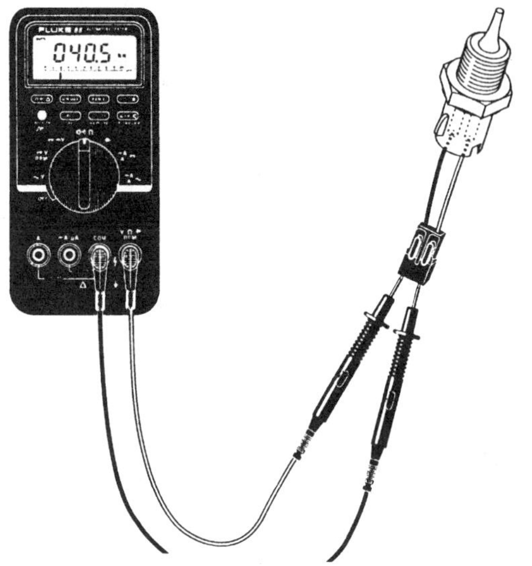

Fig. 5-20
Fluke model 80TK temperature sensor with an 80PK-6A piercing-type probe

dirty or corroded connection, a shorted wire, or possibly (last resort) an ECM problem.

Temperature Test for Ford and Other Systems which do not Provide Serial Data Output that can be Read with a Scan Tool

Note: You will need to measure the resistance of the temperature sensor *at the ECM harness connector* so that your reading includes any possible wiring problems. You will also need to measure the temperature at the base of the sensor.

1. Set up the meter to measure ohms by turning the rotary dial to W.
2. Disconnect the ECM from the harness connector and measure the resistance at the appropriate pins using a breakout box.
3. Measure the temperature at the base of the sensor and compare your findings within Table 5-2 or manufacturer's specs.
4. If your resistance measurements do not agree with the manufacturer's specs, unplug the sensor connector and measure the resistance of the sensor directly.
5. If the resistance is now within spec look for a wiring problem. Dirty or corroded terminals will cause higher-than-normal resistance, shorts to ground will cause lower-than-normal readings.

System/Component	Measurement Type				
	Voltage Presence & Level	Voltage Drop	Current	Resistance	Temperature
Charging System					
Alternators	●		●		
Regulators	●				
Diodes					
Connectors	●	●		●	
Starting System					
Batteries	●	●			
Starters		●	●		
Solenoids	●	●			
Connectors		●		●	
Interlocks	●				
Ignition System					
Coils	●			●	
Connectors	●	●		●	
Condensors				●	
Contact Set (points)	●			●	
Distributor Caps				●	
Plug Wires				●	
Rotors				●	
Magnetic Pick-up	●			●	
Lighting & Accessories					
A/C Condensors					●
A/C Evaporators					●
Compressor Clutches	●		●		
Lighting Ciircuits	●	●	●	●	
Relays	●	●		●	
Transmissions					
Cooling System					
Connectors	●	●		●	
Fan Motors	●		●		
Relays	●	●			
Temperature Switches	●	●		●	●
Radiators					●

Table 5-2
Electrical system component vs. measurement type

Throttle Position Sensors

Resistance Test

Key-Off-Engine-Off

1. Disconnect the throttle position sensor's electrical connector.
2. Connect the DMM to the sensor. Set the DMM to measure resistance by turning the rotary dial to the resistance scale.
3. With the engine not running, slowly open the throttle while watching the DMM display.
4. The analog pointer should move smoothly without jumps or skips up or down. If it does jump or skip there is a bad spot in the sensor.

Note: the analog pointer may up-range and wrap around while you are performing the test. This does not indicate a bad spot even though the display shifts to the higher range.

Voltage Test

Key-On-Engine-Off

1. Install jumper wires between the throttle position sensor electrical connectors and harness.
2. Set the meter to read DC volts by turning the rotary knob to V/RPM.
3. Connect the black lead to a good clean ground.
4. With the red lead, probe the jumper on the throttle position sensor electrical connector that provides signal return to the ECM.
5. With the key on and engine off, slowly open the throttle while watching the DMM display.
6. As with the resistance test the analog pointer should move smoothly.
7. Let the throttle return to the idle position and observe the voltage. Compare the reading with the manufacturer's specification. Adjust as necessary.

Note: Some are not adjustable and some are on/off switches, not potentiometers.

Troubleshooting Nonsensor/Computer-Related Electrical Faults

Thus far we have concentrated on relatively new cars with types of sensors older cars never had, nor did they obviously have an on-board computer. So let's assume you own an older car and we'll start with the basics. Since we are still using a DMM let's start thinking about voltage and ask ourselves three questions:

1. Is voltage present?
2. What is the voltage reading?
3. What is the voltage drop across a component?

Keeping this in mind we next need to realize automotive electrical faults are divided into several different categories, according to the system causing the trouble. We are going to examine five different areas:

1. The charging system
2. The starting system
3. The ignition system
4. The cooling system
5. Current drains from shorts and bad grounds (this is most often lighting and accessories).

Batteries

If you have a "no-start" problem the battery probably has become discharged and won't crank the engine. A fully charged battery should show at least 12.4 volts. More modern batteries are maintenance free, which means you can't get inside and take a hygrometer to measure the acidity of the electrolyte. Therefore, you'll just have to place the battery under a load and measure its terminal voltage (see Fig. 5-21). If you have re-

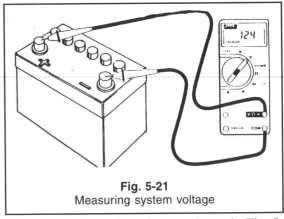

Fig. 5-21
Measuring system voltage

moved the battery, there are commercially available loads such as the one shown in Fig. 5-22. This load consists of a stack of common discs which act as seriesed low ohm resistors. These are called carbon piles.

Fig. 5-22
Typical carbon pile load

The Alternator should maintain enough voltage to keep the battery charged even while running at its rated maximum current. Check the alternator output voltage at the battery terminals with the alternator loaded to its rated output. You can also check the alternator using a clamp on current clamp (see Fig. 5-23 clamp "A"). If you suspect poor (worn) brushes you can perform a field test (see Fig. 5-23 clamp "B"). If you get a low reading of field current or the alternator output is low, the brushes are likely worn or the regulator is faulty. Shorted diodes can run the battery down overnight. Shorted diodes can be found by using the diode test function on a DMM. First, disconnect the battery wire from the alternator's output and touch one test probe to the output terminal and the other probe to the alternator housing, see Fig. 5-24. Reverse the probes and perform this test again.

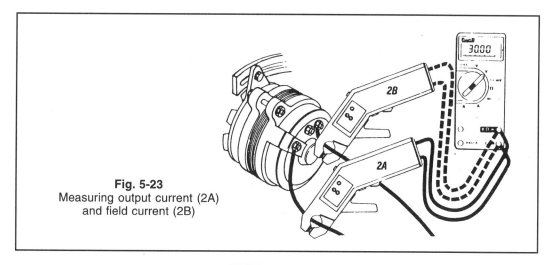

Fig. 5-23
Measuring output current (2A)
and field current (2B)

You should hear a "chirp" and the display should indicate about 0.8 volts; if it indicates about 0.4 volts, one of the diodes is shorted. If you hear a continuous tone, two diodes are shorted.

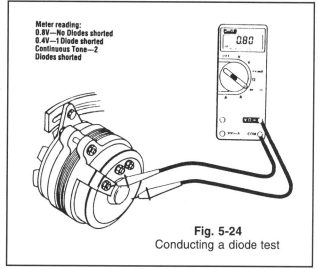

Meter reading:
0.8V—No Diodes shorted
0.4V—1 Diode shorted
Continuous Tone—2
Diodes shorted

Fig. 5-24
Conducting a diode test

Voltage Regulators

Although most cars are equipped with solid-state regulators which are not adjustable, there are some older relay type regulators which are adjustable. Low output voltage may be the alternator or the regulator, but which is at fault? Begin by bypassing the regulator to check whether the alternator is faulty. CAUTION: Use a rheostat to control field current instead of shorting the field directly to the battery. With the battery fully charged, adjust the regulator so that its output voltage matches the battery's specifications. A voltage drop test determines wires between the alternator and regulator which are almost broken, corroded terminals etc. To determine which diodes are faulty, dismantle the alternator, disconnect each diode, and check them individually.

Starting Systems

Many dead batteries have been replaced when the real fault was the starting system. If the battery passes the load test, look for resistance in the starter circuit, especially if the engine still cranks slowly.

Starter Current

Investigate excessive current draw, worn-through insulation, a seized or tight engine, or faulty starter. If the starter turns the engine slowly, the current drawn is not excessively high, and the battery is in good condition, check the resistance in the starter circuit (see Fig. 5-25). By using an inductive current clamp such as the Fluke Y8100, which is good for 200

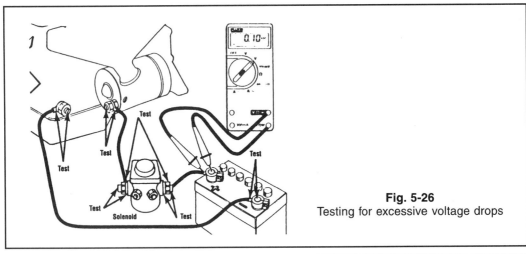

Fig. 5-26
Testing for excessive voltage drops

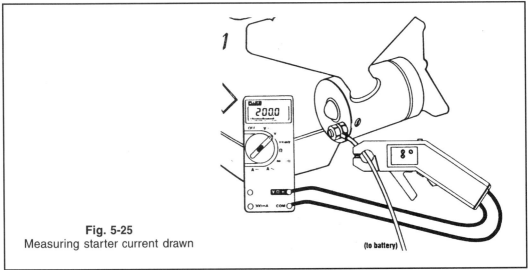

Fig. 5-25
Measuring starter current drawn

(to battery)

amps, you can measure starter current. Even very low resistance in the starter circuit will cause it to turn the engine over slowly. As an example, at 200 amps, a 0.01 ohm resistance will cause a 2-volt drop! A resistance of 0.01 ohms, though, is beyond all but the most sophisticated DMMs to measure. So, by measuring the voltage drop, you can tell where the excessive resistance exists (see Fig. 5-26). Measure voltage drops across each component and connection in the starter circuit while cranking your engine. Measure the voltage drop between the battery post and the connecting cable, the solenoid posts and the wires that attach to them, and across the solenoid itself. Also, check the connection on the starter, and the ground strap connection to the engine block.

Voltage Drops

There is a TouchHold function on certain Fluke DMMs, which allows you to record the voltage drop while you are inside car cranking it. When you come out to change test points it retains the readings. Since it ignores zero readings there is no confusion here. Typically, see Fig. 5-27, you would engage the TouchHold function and place the DMM into the mV range. Place the leads across the connection to be measured, crank the engine from 4 to 5 seconds and these readings will be held and displayed on the DMM.

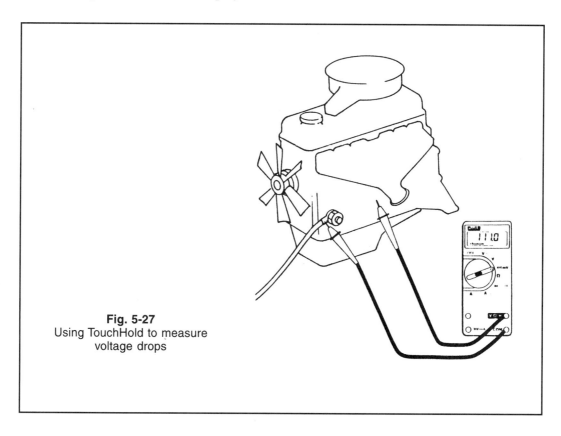

Fig. 5-27
Using TouchHold to measure
voltage drops

Ignition System

DMMs will measure down to a tenth of an ohm while analog meters will rarely measure below an ohm. When measuring ignition coils (see Fig. 5-28), check the resistance of the primary and secondary windings. Perform this test while the coil is both hot and cold. Faulty coils should have a primary winding with a low internal resistance and a secondary winding with a higher internal resistance.

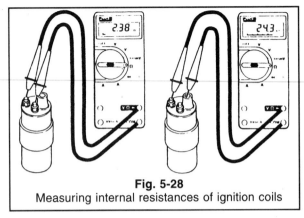

Fig. 5-28
Measuring internal resistances of ignition coils

Condensers

These are capacitors, but they were also called condensers 20 years ago and have retained that term is auto parlance. You can check for a leaky capacitor with the ohms scale by having the DMM charge the capacitor, where the resistance should increase to infinity (see Fig. 5-29). Any other readings indicate a faulty capacitor (condenser). If the condenser is on the car, be certain the points are open.

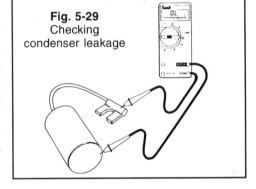

Fig. 5-29
Checking condenser leakage

Spark Plug Wires

You should check spark plug wires if they are more than two years old or if your scope indicates a problem. Measure resistance by placing a probe from the DMM on each end (see Fig. 5-30). The reading should be on the order of 1 kilohm per foot of length; although some high-energy ignition spark plug wires have 30,000 ohms per foot.

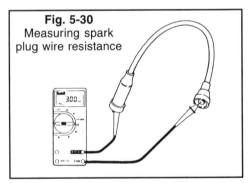

Fig. 5-30
Measuring spark plug wire resistance

Rotors

Corrosion or arcing at the rotor's tip can be a problem. Also, check for high resistance between the center contact and the rotor tip. Remember, some foreign cars have rotors with rev limiters built in. Start by disconnecting the distributor from the ignition module. Connect the DMM across the pickup and set it to DC volts. When the engine is cranked, pulses should appear on the DMM's bar graph. If no pulses appear it indicates that the reluctor wheel or the inductive pickup is bad.

High-Energy Ignitions

Since the spark plug wires in these systems have such high resistance, as just stated, it is often difficult to determine which of the components in an electronic ignition system is faulty. By process of elimination you can isolate the trouble to either the electronic module or the distributor's inductive pickup. Check the proper operation of the inductive pickup and reluctor wheel by measuring the output pulses as we did in Fig. 5-31. However, on Chrysler and other Hall-effect ignitions, this can't be done.

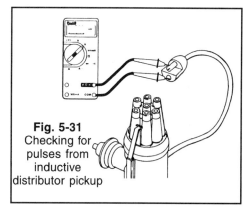

Fig. 5-31
Checking for pulses from inductive distributor pickup

Current Drains, Shorts, and Bad Grounds

The cause of these problems is perplexing since it often appears to have nothing to do with the symptom. Current drains that run the battery down are often not dead shorts, but shorts of higher-than-normal resistance. Let's start with the alternator leakage current test. Here, you connect the DMM in series with the alternator output terminal while the car is *not* running. Leakage current should be on the order of a few milliamps maximum, and more often will be on the order of 500 microamps (see Fig. 5-32).

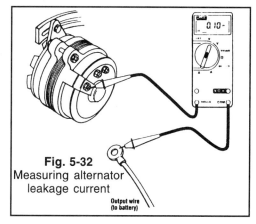

Fig. 5-32
Measuring alternator leakage current

Output wire (to battery)

Locating Current Drains

In order to avoid blowing your DMM's internal fuse, use the 10 Amp scale until you have safely determined the current is below 320 mA. Do not crank the engine or run accessories which draw more than 10 Amps. After determining the current is less than 300 mA, switch over to the 320 mA scale and place one lead in this terminal and the other lead is the COM terminal. You can isolate the circuit causing the current drain by pulling one fuse after the other (see Fig. 5-33) until zero amps are drawn. Next, check the leakage current of the alternator as you did in Fig. 5-33. Once you know how much current the alter-

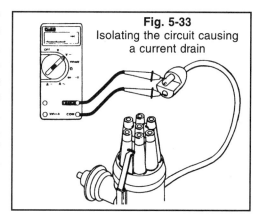

Fig. 5-33
Isolating the circuit causing a current drain

nator diodes draw you can leave it disconnected to avoid confusing its current draw with the culprit. Also, under-hood and trunk lights, dome lights, and computers all draw current. A piece of tape over the door switch can prevent this light from coming on.

Shorts are usually caused by a defective component or insulation which has rubbed through. The old practice of using the DMM in series on the volts scale hardly works any longer due to on-board computers, and even in older cars there are still many digital circuits, such as car clocks. All of these draw current. So, if you are working on a noncomputer car, use the same process as we did earlier in switching out fuses; how-

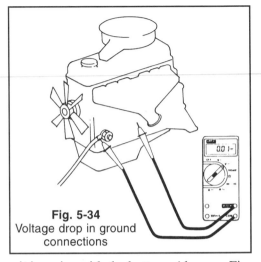

Fig. 5-34
Voltage drop in ground connections

ever, place the DMM onto the volts scale and place it is series with the battery. Also, see Fig. 5-34 and realize you can find a bad ground by checking the voltage between the component's ground wire and a clean chassis ground or the negative terminal. Voltage readings higher than a few hundredths of a volt mean there is a corroded terminal that should be cleaned.

Dedicated Testers

Throughout this chapter we have used, to this point, a DMM to troubleshoot our cars' electrical systems and even the on-board computer. But as you already know, to be fair, we must mention the existence of some commercially available diagnostic systems which cost about $1,000. They are designed solely for a car with an on-board computer. These on-board computers have a series of inputs and outputs. The inputs come from a network of sensors distributed throughout the car (mostly in the engine compartment and the dash). The computer interprets these inputs and sends outputs or commands to the car's on-board computer.

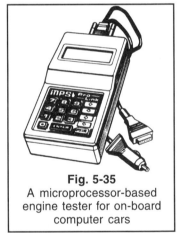

Fig. 5-35
A microprocessor-based engine tester for on-board computer cars

When something goes wrong with the car's operation, it may be due to a faulty input, a faulty command, or a faulty reaction to a command. MIPS (Micro Processor Systems, Inc.) of Sterling Heights,

Fig. 5-36
"Intelligent" breakout box accessory for the microprocessor-based engine tester.

Michigan, makes a microprocessor-based test instrument designed to tap into these on-board computers (see Fig. 5-35 and note the cigarette lighter power input and the connector on the other end). There are some sophisticated accessories to this instrument, such as the display that is an "intelligent breakout box" (see Fig. 5-36). This is for a Ford EECIV engine which was what we troubleshot here. The photo says Toyota, but the modules (programmed memories) adapt it to another generic series of engines as well. Next, we will show you how to use the Fluke model 97 DSO to diagnose and troubleshoot some very specific examples of typical trouble spots within your car's electrical system.

Making Rotational Measurements

The Fluke PM 3384E is a powerful 100 MHz, 200 Mega Samples per second digital storage oscilloscope combined with a full-featured analog oscilloscope. It is equipped with a new time base able to adapt automatically to changes in signal frequency. This eliminates the need for constant manual adjustment of the time base. This feature makes it ideal for measuring speeds in rpm of rotational objects, such as cam and crankshafts, or other tasks involving variable speed rotating machines such as fuel injection systems and variable speed motor drives.

Digital Variable Time Base

The Fluke PM 3384E has a new variable time base when operating in digital storage mode. This is the first time this feature has been included in a CombiScope™. This function is extremely useful when placing a signal over a desired number of divisions. Also, switch-ing between analog and digital storage modes to make a hard copy now gives a consistent display.

Time-Base Readout in Degrees per Division

Repetitive waveforms are all based on cycles of 360 de-grees. These waveforms can be from an engine, as shown above left, or any other con-tinuous signal source. Specify-ing the time base in signal de-grees per division means that the Fluke PM 3384E automati-cally adjusts its time base

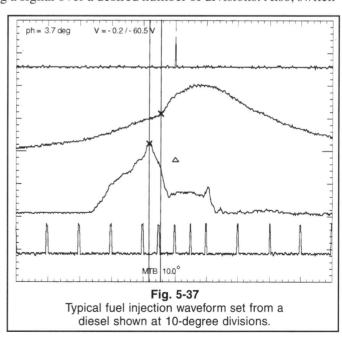

Fig. 5-37
Typical fuel injection waveform set from a
diesel shown at 10-degree divisions.

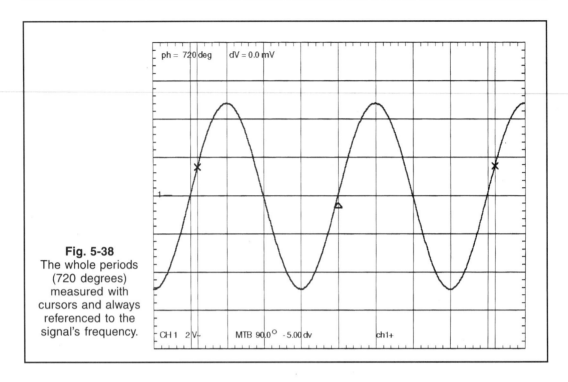

ph = 720 deg   dV = 0.0 mV

Fig. 5-38
The whole periods
(720 degrees)
measured with
cursors and always
referenced to the
signal's frequency.

CH 1   2 V   MTB 90.0° -5.00 dv   ch1+

speed, following frequency changes, to always show the same signal display. So signals in applications as diverse as internal combustion engines changing speed to phase-locked loops can all be displayed without adjustment to the time base control (see Fig. 5-37).

Autoranging Time Base for Hands-Free Operation

When used in the normal time per division mode, the Fluke PM 3384E can be set to autorange on the input signal frequency. This is ideal for applications such as frequency response testing, voltage controlled oscillators, and variable frequency transducers because the oscilloscope does not need to be adjusted when the signal frequency changes. Autoranging works in the normal 1-2-5 step sequence and in fully variable mode. You simply specify how many waveform cycles you want to see on the screen and the Fluke PM 3384E does the rest automatically (see Fig. 5-38).

Automatic Phase Measuring Cursors

The cursor operation for phase measurements on the Fluke PM 3384E reaches new levels of performance and ease of use. The cursors are automatically referenced to the signal frequency and period giving instant results accurately and without having to manually set the 360-degree reference. Applications such as thyristor and SCR power controllers and engine crank angle-based measurements are easy using this feature. The cursors can also be set to measure between channels to give phase shift, perfect for working on variable phase signals as found in positional control (servo) systems.

Automatic Channel-to-Channel Delay and Phase Shift Measurement

The direct measurement facilities have been extended to include automatic channel-to-channel delay and phase shift measurement. When operating in the time-base degrees-per-division mode the results are given in degrees, ideal for checking the phase shift through amplifiers or filters. When operating in the time per division mode, the result is given in seconds for applications such as a group delay measurement or a propagation time measurement.

Special Features for Engine Testing

The Fluke PM 3384E automatic time base can be selected for operation on two- or four-stroke engines. One engine cycle of a four-stroke engine is comprised of two engine revolutions, i.e., 720 degrees of crankshaft rotation. When set for four-stroke operation, half engine speed signals such a needle lift that ignition pulses, and fuel pump signals can still be used as a source for the automatic time base. All measurements and readouts are rescaled by a factor of two, converting results to a crank angle base. The benefits of the automatic time base and measurement functions can now be enjoyed without the user having to apply timing correction factors. Direct measurement of rpm is now also standard. The amplitude readout can be scaled in pressure units for use via a suitable transducer and preconditioner. The 8K per channel memory depth ensures sampling at 0.1 degree resolution over more than two engine revolutions (see Fig. 5-39).

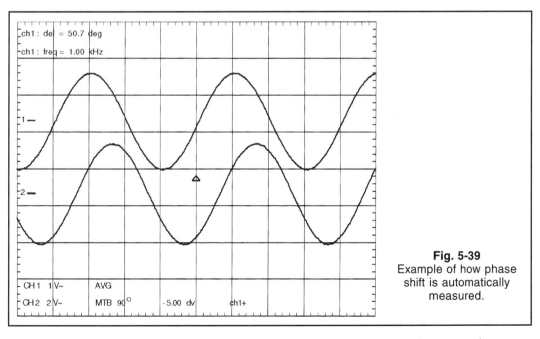

Fig. 5-39
Example of how phase shift is automatically measured.

Guidelines for Engine Signal Monitoring with the Fluke PM 3384E

The Fluke PM 3384E is a unique version of the PM 3384 CombiScope™ instrument, which is a combined fully featured analog and digital storage oscilloscope (DSO). Thanks to its true variable sample rate this, "E" model has new functions making it especially suitable for monitoring signals associated with internal combustion engines. Features include:

- Automatic time base operating in degrees per division.
- Automatic time base scaling for two- or four-stroke engines.
- Automatic rpm measurement.
- Channel scaling in pressure and other engineering units.
- Cursors with a direct measurement of degrees, independent or engine rpm or input signal frequency.
- FFT analysis for "pinging" and vibration detection.

This portion of Chapter 5 covers the display of waveforms from reciprocating internal combustion engines. There are other types of engines such as the rotary or wankel engine and gas turbine engines. There are many other types of rotating machinery. The PM 3384E is also suitable for monitoring changing frequency and angular displacement signals from these sources.

Measurements on engines are based on the angular position of the crankshaft. This is the basis for the timing of all other engine parameters. The measurements are made relative to Top Dead Center, TDC. Typical signals monitored during engine testing are cylinder pressure, fuel injector needle lift, fuel line pressure, and timing between various events during engine operation. Most engine test systems also provide a timing marker signal that is derived from the crankshaft. It consists of a pulse usually at TDC sometimes combined with other pulses at certain degree intervals. The timing markers are often displayed as a reference together with the other engine signals, as shown later in Fig. 5-43. By scaling the time base in degrees per division and changing the time base sweep speed automatically to follow changes in engine rpm, typical measurements such as ignition delay and start of needle lift can be made directly using the cursors.

Compatibility with Older Systems

Older engine test systems use four-channel analog oscilloscopes, often triggered via the external trigger input. The PM 3384E is the ideal plug-in replacement in these applications. The PM 3384E has an external trigger input fitted as standard in addition to its four full channels. It also operates as an analog oscilloscope so continuity in working practices and test procedures can be maintained. In digital storage mode PM 3384E has a manually adjustable variable time base, just like in analog mode.

PM 3384E Time Base Modes

There are four time base modes available (see Table 5-4). All are fully calibrated all the time. All measurements retain their full accuracy in each mode whether made automatically or by cursors.

Timebase Mode	Timebase Units Mode	Analog Mode	Digital Storage	Notes
Manual 1-2-5 steps	time / division	yes	yes	
Manual variable	time / division	yes	yes	usually not found on DSOs
Auto-Ranging 1-2-5	time / division	no	yes	
Auto-Ranging Variable	time / division	no	yes	
Degrees Mode Normal	degrees/ division	no	yes	
Degrees Mode 4-Stroke	degrees/ division	no	yes	All x-axis readouts are multiplied by 2. Shown by an asterisk beside the timebase readout.

Table 5-4
Time base modes and units of measure

Four-Stroke Engines—Where to Trigger

As in all oscilloscope applications, triggering is the secret to successful signal viewing. The strokes in a four-stroke engine cycle are namely:

1. Intake
2. Compression
3. Power
4. Exhaust

Ignition occurs between compression and power strokes and can be spontaneous as in a diesel engine or spark-initiated as in a gasoline engine. In the intake and power strokes the piston moves down the cylinder. In the compression and exhaust strokes it moves up the cylinder; therefore, one engine cycle comprises of two crankshaft revolutions. Most existing monitoring and test systems trigger from a pulse derived from the crankshaft (see Fig. 5-40). Because there is a pulse every revolution there is no easy method to select the pulse from the compression/power revolution, usually the one of interest. One method is to trigger from signals from the combustion process that occur every engine cycle, e.g., fuel injector needle lift or cylinder pressure. The frequency of these signals is the same as that of the camshaft rotation. "FOUR-STROKE" must be selected in "AUTO-TB CONFIG" menu when triggering from such signals. The "AUTO-TB CONFIG" menu is accessed through the "ACQUIRE" button.

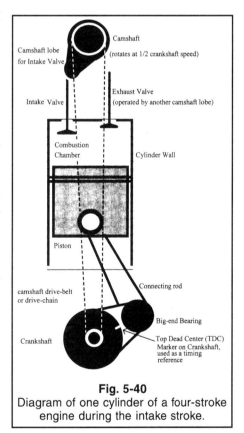

Fig. 5-40
Diagram of one cylinder of a four-stroke engine during the intake stroke.

When "FOUR-STROKE" is selected and the degrees mode is active, all degree-based timing measurements, cursor readouts, and the time base degrees readout are multiplied by

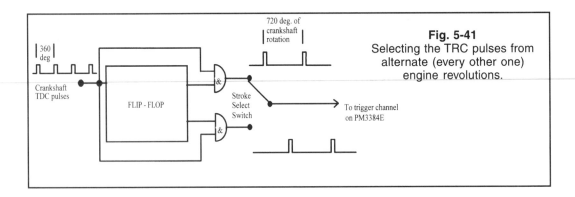

Fig. 5-41
Selecting the TRC pulses from alternate (every other one) engine revolutions.

two because 360 degrees of camshaft rotation, or one engine cycle, equals 720 degrees of crankshaft rotation.

The best measurement results and ease-of-use benefits are obtained when the PM 3384E is triggered from a pulse at TDC between the compression and the power strokes. Some engine test cells have such a TDC pulse available for use with other equipment. If such a pulse is not available it can be easily obtained by constructing the setup in Fig. 5-41. Once the engine is running simply toggle the switch to select the correct stroke. The AND gates preserve the width of the TDC pulse. Use the flip-flop outputs directly if you do not want to preserve the pulse width. Doing this will produce a square wave at camshaft rpm. This can help with stability when using the time base at higher degrees-per-division settings and the trigger pulse width is narrow.

Two-Stroke Engines

In a two-stroke engine the complete engine cycle takes place during one revolution of the crankshaft. It is therefore very easy to use the PM 3384E on these engines. The engine can simply be considered as a function generator so you can trigger from any convenient signal. As with the four-stroke engine a pulse at TDC will give you the most benefit from the PM 3384E. Ensure that "NORMAL" is selected in the "AUTO-TB CONFIG" submenu when using the PM 3384E on two-stroke engines.

Memory Length (samples)	Auto-TB range (deg/div) normal mode	Max. Resolution (degrees) normal mode	Auto-TB range (deg/div) Four-stroke	Max. Resolution (degrees) speed Four-stroke	Screen update speed
512	90 to 360	1.8	180 to 720	3.6	Fast
4K	10 to 360	0.2	20 to 720	0.4	Medium
8K	5 to 360	0.1	10 to 720	0.2	Slower

Table 5-5
Resolution and time base range vs. memory length.

Resolution vs. Memory Depth

The PM 3384E has three different user-selectable memory lengths for automatic degrees-per-division operation. These are: 512 samples, 4K samples, and 8K samples for each channel. They offer a balance between update speed and horizontal resolution. In 512 sample mode the time-base range is 90 to 360 degrees/division. Each horizontal division has 50 samples displayed when no magnification is used, so the maximum horizontal resolution is 90 degrees/ 50 samples or 1.8 degrees between samples. This mode also gives the fastest screen update. In 4K sample mode the range is 10 to 360 degrees/division. This gives 0.2 degree resolution and medium update speed. The highest resolution is in the 8K samples mode, where the range is 5 to 360 degrees/division giving 0.1 degree resolution. In "FOUR-STROKE" mode the crank angle measurement resolution is halved and the readouts are doubled. This means that in 8K samples mode the fastest time-base speed is 10 degrees/division giving the highest resolution of 0.2 degrees of crank angle between samples. The screen update is very dependent on the engine rpm. These relationships are summarized in Table 5-5.

Trigger Pulse Width

To ensure that the PM 3384E Auto-TB is stable, the trigger pulse must be wide enough to be sampled several times during each revolution. The PM 3384E needs these samples to calculate the signal period and to set the sample clock rate to give the correct number of degrees per division. When the trigger pulse is not wide enough to be sampled several times, then the automatic time base will be unstable.

There are 50 samples per division when no magnification is used. The rule of thumb to check if a pulse is wide enough is: Minimum trigger pulse width = 2.5/50 ÷ deg/div setting = 0.05 x deg/div setting.

For example: PM 3384E set to 5 deg/ div. Minimum trigger signal pulse width = 5 x 0.05=0.25 degrees.

Using the Cursors in Degrees Mode

When the PM 3384E is operating in the automatic time base degrees mode, the cursors use the trigger point as the zero degree reference. The cursors are also automatically referenced to the trigger signal 360-

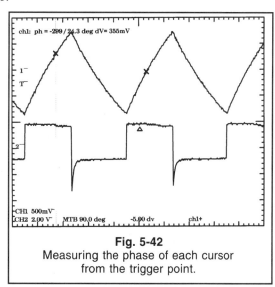

Fig. 5-42
Measuring the phase of each cursor
from the trigger point.

degree period when the degrees mode of the Auto-TB is active and cursor 1 is assigned to the trigger channel. This means that regardless of input signal frequency or changes in it, direct degree measurements can be made. The "X" in Fig. 5-42 at the intersections of the cursors and trace indicates that both cursors are active on Channel 1. There are two degree cursor modes, namely "Phase 1 and Phase 2" and "Phase difference." In Phase 1 and Phase 2 mode the readout gives the degrees each cursor is from the trigger point. Figure 5-42 shows the first cursor at 299 degrees before the trigger point and cursor 2 at 24.3 degrees after the trigger. If the trigger pulse is at TDC then the Ph1 and Ph2 mode will give direct readout of the cursors' position with respect to TDC.

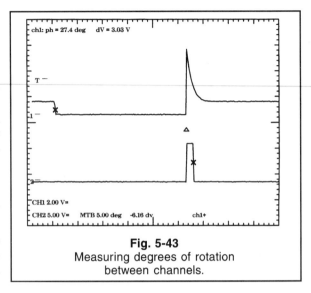

Fig. 5-43
Measuring degrees of rotation
between channels.

The "Phase difference" mode gives the distance in degrees between the cursors. Figure 5-43 shows cursor 1 operating on the trigger channel, channel 1. This gives the automatic cursor phase referencing. Cursor 2 is operating on Channel 2. The phase readout is the number of Channel 1 degrees separating the two cursors. The voltage measurement is the voltage between cursor 1 and cursor 2.

Note: If operating in degrees mode, cursor 1 is moved to a channel other than the trigger channel, then the phase readouts will switch back to time readouts. This avoids the chance that an incorrect degrees result is obtained because the signal on the other channel has a different frequency.

Switching cursor 1 back to the trigger channel restores the degrees readout and automatic degrees referencing. If cursor 2 is moved to another channel the degrees readout remains referenced to channel 1 as in Fig. 5-43. To measure degrees on a channel other than the trigger channel the cursor 360-degree reference must be set manually as in normal time base modes.

Using the Cursors when the Trigger is not at TDC

If the trigger pulse is not at TDC but is at a known fixed offset from TDC, for example 70 degrees, then the cursors can still be used to make direct measurements relative to TDC. First of all, locate in the Ph1 and Ph2 mode cursor 1 so that its readout is the same as the known trigger pulse offset from TDC, in this example at 70 degrees. Now change the cursor mode to "Phase difference" and use cursor 2 (the delta cursor) to measure to the desired point on the wave form. The cursor readout will be the degree difference between the cursors, which is the same as the number of degrees that cursor 2 is from TDC. These two setups can be stored in the PM 3384E's setup memories and be quickly recalled when required. When performing a repetitive series of tests, operation can be made even faster by reprogramming the Autoset button, via the "UTILITY" menu, to sequence through the stored instrument setups.

Pressure Units Readout

With the PM 3384E it is possible to change the channel units for each channel individually from volts to other physical and electrical units. One of these is "P" for pressure. If your pressure measuring system gives, for example, 10 mV per PSI, then in the "PROBE COR-RECTION" menu under the "UTILITY" key you can change the units to "P" and apply a correction factor of 100. All cursor and automatic measurements on the rescaled channel will now give a direct readout of your pressure units.

Working with Higher Frequency or rpm Signals

The Auto-Time Base is digitally controlled; it changes sweep speed in steps. The step size is not constant over the range of sweep speeds. At low frequencies/rpm these steps are very small compared to the signal period so the time base can change its sweep speed in small steps to respond to changes in signal frequency. At higher frequencies however, usually beyond the range of engine signals, the time base step size becomes significant when compared to the signal period. For example, further up the range of sweep speed the step size is one microsecond. This means that a frequency change of up to one megahertz is required before the Auto-TB adjusts itself to the new frequency. Obviously, any fixed degrees-per-division readout for the time base could be very wrong before the Auto-TB readjusts. To overcome this error possibility PM 3384E detects that the demanded degrees-per-division and actual degrees-per-division differ and it displays the actual degrees-per-division as the time-base readout. The demanded value is still available in the "ACQUIRE" menu. None of this, however, affects the cursor readouts or automatic measurement results. These maintain their accuracy and resolution at all times because they are linked to the crystal controlled sample clock.

Documentation and Data Storage

To make hard copies of the screen, PM 3384E has an RS 232 serial interface fitted as standard. The PM 3384E will give you a what-you-see-is-what-you-get (WYSIWYG) picture on laser and dot matrix printers. It will also output to a HPGL-compatible digital plotter.

To import waveforms into a PC for documentation as in Fig. 5-44, Fluke offers you an optional software package, DSO COM. This package can acquire data and setup information from the PM 3384E in formats suitable for inclusion in word processing packages, spreadsheets, and graphics packages where text can be

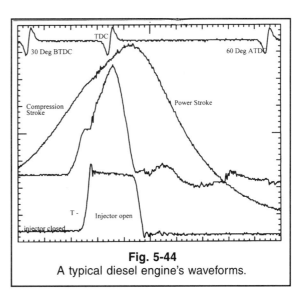

Fig. 5-44
A typical diesel engine's waveforms.

added as in Fig. 5-44. The waveforms and setup information can be stored on disk and sent back to the PM 3384E later for comparison purposes.

Further Analysis

The Fast Fourier Transform or FFT function is a mathematical process that analyzes a signal and displays its different frequency components. When the frequency composition of a signal changes as a result of, for example, knocking (pinging) or increased vibration, the change will be much more readily seen in the FFT display than in the normal waveform display.

Chapter Quiz 5

True or False

1. The vast improvement in gas mileage on modern cars is mainly due to your car's computer making decisions based upon the data it is supplied by all of your car's sensors.
2. In current cars, airflow and its mass, as it exists in your car's exhaust manifold, is measured directly by electronic sensors.
3. A modern multiport fuel injection system precisely dispenses gas vapor and times the spark.
4. Your car's computer can suffer from inaccuracies because of the time lag between the instance it receives data and the amount of time it takes to fully process this data.
5. There are mechanical sensors that directly measure or gage airflow.
6. One new type of sensor is based on a single silicon piezoelectric element producing a voltage that is inversely proportional to the magnitude of the pressure it senses.
7. An exhaust analyzer can not tell you if you are experiencing a too rich or to lean condition.
8. It will cause no harm if you use an analog multimeter to perform a resistance test on your car's computer.
9. By monitoring your EGO sensor's output voltage you can determine if your engine is receiving excessive gas fumes.
10. You can check to see if there are any opens in your EGO circuit when you have this sensor still in place (without unhooking it).

Matching

11. You do not want to remove these while your ignition key is in the on position.
12. Your car's oxygen sensor sends data to the onboard computer to control this.
13. These types of sensors actually act as small batteries.
14. A rich condition means there is too much fuel present and too little quantity of this.
15. A properly functioning oxygen sensor oscillates back and forth between these two states.
16. When your oxygen sensor becomes dirty, this is obvious because its output voltage swings do this.
17. This sensor determines your engine's rpm.
18. An air charge temperature sensor determines what condition in your car's operating environment.
19. The AC-Delco MAF sensor is a form of a "hot wire anemometer" and produces this form of electrical output.
20. Wires used in most car sensors possess this characteristic that relates their temperature change to their resulting resistance.
21. These tell you the position of your camshaft and crankshaft.
22. The sensors in question 21 above come in these two types.
23. Hot, dry air has this characteristic, with respect to hot moist air.
24. This produces a 5-volt square wave each time the vane wheel of this type of sensor cuts through the magnetic field supplied by this sensor's permanent magnet.
25. This is a synonym for magnetic sensor.
26. The type of sensor referenced in question 25 produces this type of output signal.
27. This type of sensor (a new different type) also works on the same principle as the sensor in question 26.
28. Your car's ECM (computer) provides a control signal turning on a transistor that in turn provides current a path to ground. When this happens, what does it open?
29. When you are driving at high speeds, and under a heavy load, your injectors are on nearly a full one of these, what is it?
30. A feedback carburetor operates and is commanded by the voltage produced by this sensor.
31. The fuel metering rod in a Rochester-type carburetor has a control solenoid that responds to this.
32. Your idle air control (IAC) motor comes in two types. One type is current driven to hold its position. What is the other type of this motor?
33. This type of idle actuator rotates only 609 degrees between being fully open and fully closed.
34. Your idle speed motor receives pulsing current and its source varies its duty cycle (on-time to off-time ratio) so that this motor really responds to what electrical component?
35. This sensor is merely a variable resistor attached to your car's throttle shaft.

A. Zirconia
B. Throttle position sensor
C. The amount of oxygen
D. Converges
E. Cam and crankshaft sensors
F. Square wave
G. Spark plug wires
H. Magnetic type ABS wheel speed sensor
I. Hall-effect sensor
J. Oxygen and no oxygen
K. Stepper motor
L. AC pulse
M. NTC (Negative Temperature Coefficient)

N. Bosch rotary idle actuator
O. Oxygen
P. Crankshaft revolution
Q. Frequency
R. The O_2 sensor
S. The injector valve
T. Magnetic and Hall effect
U. Relative humidity
V. Injector ON-time
W. Variable reluctance sensor
Y. Less dense
Z. Crankshaft and position sensor

6

More Troubleshooting
Applications with a DSO

This very liberally illustrated chapter examines actual troubleshooting procedures, and the waveforms you can expect to see when determining various problems associated with your car's electrical system. The sole instrument of choice used in Chapter 6 is a DSO. We will briefly use a TekMeter, which is a DSO and DMM combined (see Fig. 6-1). But our main focus will be the Fluke 97, which we have already covered, as well as the first DSO we will profile, the Tektronix ultra-portable 222A. The Tek 222A DSO's real value lies in its small size, rugged construction, single pushbutton setup, and uncluttered front panel with a minimum of unconfusing controls. This is ideal for many mechanics who are not intimately familiar with electronics to any in-depth degree.

In this portion of Chapter 6, you will learn to recognize types of automotive circuits and their respective waveforms. In addition, specific examples of waveforms created while testing various automotive components will be included. Finally, you will learn how to connect the Tektronix 222A DSO to automotive circuits.

Getting Started

There are three simple steps to begin:

1. Select the ON button; the 222A turns off automatically when not being used.
2. Attach the probes to the automotive component to be tested. More detailed instructions, related to specific applications, are included in the body of this manual.
3. Push the Auto Setup button which prepares the 222A to acquire and display signals.

Circuits commonly found on automobiles can be classified by category. By looking at circuit types and learning about corresponding waveforms, you will be able to predict what waveforms will appear during tests of various automotive components and systems. As a

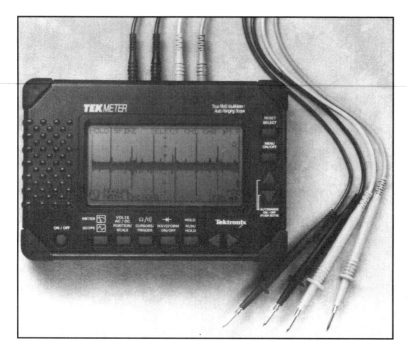

Fig. 6-1
The TekMeter, which is
a DSO and DMM
combined.

review of signal shapes that we have already lightly covered, let's revisit them in a more practical fashion now.

Sine Waves

A pure sine wave is produced by an AC (alternating current) generator. The characteristic waveform is created because of the angle at which the armature windings pass through the magnetic field. It changes and then reverses direction according to a trigonometric function known as the *sine* function. If you graph the sine function you will see a sine wave just like the scope pattern produced by an AC generator.

Some examples of devices that produce sine waves:

- Power company generator
- Permanent magnetic type vehicle speed sensors
- Alternators (before the output is rectified by the diodes)

These devices produce a sine wave because the angle of a rotating part changes with relation to a stationary part according to the sine function. There are also devices that generate their own electricity and produce a waveform somewhat like a sine wave. Some examples are:

- Variable reluctance crankshaft sensors
- Anti-lock brakes
- System wheel sensors

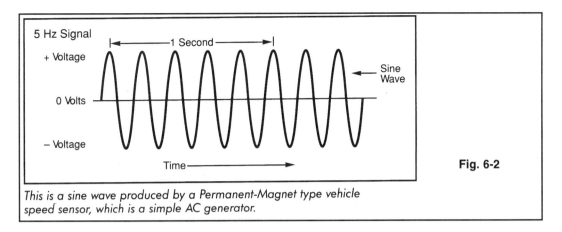

This is a sine wave produced by a Permanent-Magnet type vehicle speed sensor, which is a simple AC generator.

Fig. 6-2

What is Frequency?

To review this subject, counting the number of complete sine waves a generator produces in one second determines frequency. The signal goes through a certain number of cycles per second. This measurement has been named Hertz (Hz) in honor of an early scientist by that name. The Hz measurement applies to any repetitive signal including square waves and pulse trains. (For example, common household current runs at 60 Hz.) (See Fig. 6-2).

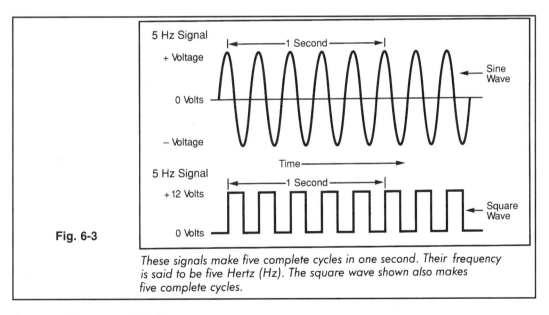

Fig. 6-3

These signals make five complete cycles in one second. Their frequency is said to be five Hertz (Hz). The square wave shown also makes five complete cycles.

Square Waves and Pulses

Computers "think" only in terms of zero or one. Square waves are signals with two voltage levels that represent these values, see Fig. 6-3. Signals that spend more time at one voltage level than another are called pulses or pulse trains.

In most automotive systems, 0V and 5V represent the numbers 0 and 1 for information coming into computer from such things as Hall switches or MAF (Mass Airflow) sensors. Voltage levels of 0V and 12V are typically the "on" and "off" levels for devices that computer controls, such as canister purge solenoids or fuel injectors.

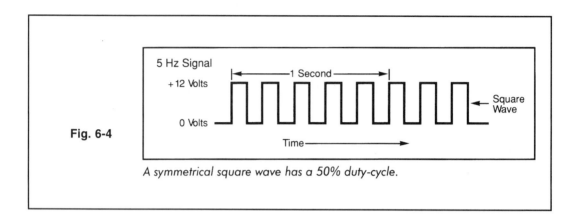

Fig. 6-4

A symmetrical square wave has a 50% duty-cycle.

Square Waves

The term square wave is often applied to any wave form that has vertical sides and a flat top. However, strictly speaking, a square wave alternates between two voltage levels giving equal time to each level. Square waves are said to be symmetrical (see Fig. 6-4). Square waves are often produced by multivibrators. Multivibrators are a type of "flip-flop" which have a square-wave output. These signals can be read by a computer.

Many square waves found in automotive electronic systems are converted from sine waves. The buffer amplifier found in GM permanent-magnet Vehicle Speed Sensor systems falls into this category.

Pulse Trains

Any signal that spends more time at one level than another is considered a pulse train. Pulse trains are said to be asymmetrical. Pulse trains send information to the computer from such devices as Hall switch-type crank position sensors, telling it when the No. 1 cylinder is at top dead center (TDC). They drive fuel injectors, controlling "on" time. And they can carry information from the computer to the ignition module, telling it when to fire the coil.

Pulse trains are seldom, if ever, made form sine waves because sine waves are symmetrical while pulse trains are not (see Figs. 6-5 and 6-6). Pulses can be generated from signals of any type by a flip-flop circuit designed for that purpose. On a GM system, the pulse the electronic control module (ECM) sends to the C31 ignition module is triggered by the signal from the variable-reluctance, crank-position sensor (see Fig. 6-5 again).

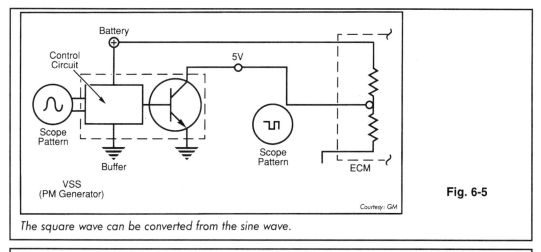

Courtesy: GM

Fig. 6-5

The square wave can be converted from the sine wave.

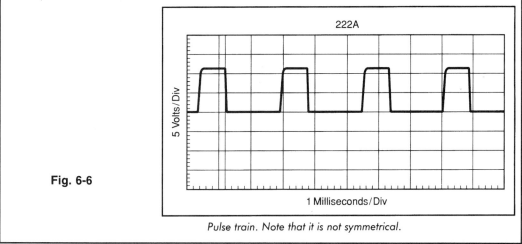

Fig. 6-6

Pulse train. Note that it is not symmetrical.

What is Pulse Width?

Pulse width is the length of time (usually in milliseconds) a pulse-type signal stays in its active state. (1000 milliseconds is equal to 1 second.) The active state is the voltage level that causes the device to be on. For example, the pulse width of a fuel injector signal indicates a low-voltage condition, since the drivers provide ground for the injector. Other components, however, may indicate a high-voltage condition when measuring pulse width. Pulse width is closely related to duty cycle, which is the percentage of time the signal stays active (see Fig. 6-7).

Pulse Width Modulation (PWM)

Changing the pulse width of a signal over a period of time to control a device is called pulse width modulation (PWM). Average current flow in any device controlled by PWM is proportional to the percentage of time the voltage is at its active level.

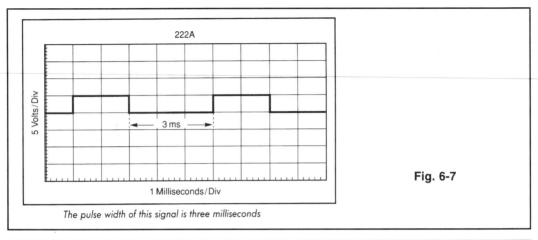

The pulse width of this signal is three milliseconds

Fig. 6-7

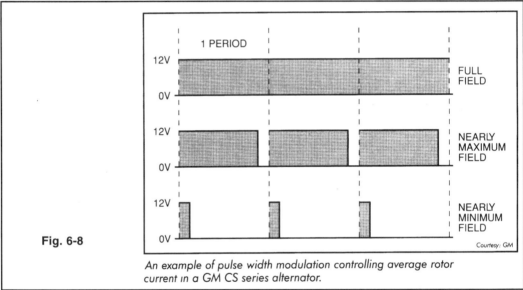

Fig. 6-8

An example of pulse width modulation controlling average rotor current in a GM CS series alternator.

For example, the regulator controls rotor current in a CS series generator by PWM. Current switches on and off at a constant frequency but the pulse width changes depending on the alternator's load. The average rotor current changes according to the pulse width (see Fig. 6-8).

Just as the output of an alternator depends on average field current, the power of a permanent-magnet DC electric motor is directly related to the average current through its windings. PWM signals can control motor speed and power. For example, some blower motors and fans are controlled by PWM.

The distance a solenoid moves against its return spring is related to the average current through its windings—the more current, the greater the motion. A PWM signal often controls a solenoid's motion. For example, PWM signals control turbo-charger waste-gates, canister-purge solenoids, and EGR (Exhaust Gas Recirculation) valves.

Effects of Inductance on Waveforms

The motors, generators, and solenoids we've discussed so far are basically inductive devices; that is, they rely on electromagnetic fields to operate. Electric current flowing in a wire creates a magnetic field that surrounds the wire. Coiling the wire, as in a motor armature, concentrates the field. The concentrated magnetic field is stored energy. When the current stops flowing, the field collapses, releasing the energy.

Any time a magnetic field moves through a conductor, such as generator winding, an electric current is produced. So, a collapsing field moving through a motor winding creates an electric current, momentarily turning the motor into a generator. The same thing happens in a solenoid, a fuel injector, or a relay.

Sometimes the voltage generated lasts only a very short time but can reach hundreds or even thousands of volts. This is the principle on which an ignition coil works. The voltage "spike" can be very damaging to solid state devices such as those found in an ECM. Other times the field collapses only partially, creating unpredictable noise voltage in electric motors whose brushes bounce and arc.

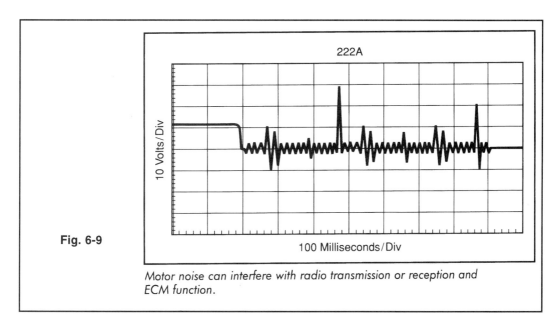

Fig. 6-9

222A

10 Volts/Div

100 Milliseconds/Div

Motor noise can interfere with radio transmission or reception and ECM function.

To prevent interference or damage, several devices are used to block noise or spikes or shunt them harmlessly to ground (see Fig. 6-10). These are capacitors, inductors, and diodes. Capacitors and inductors are unique in that they react differently to AC than to DC. Capacitors are two layers of foil, separated by an insulator that absorbs electrons, releasing them later. Capacitors will block DC while passing AC. Inductors, on the other hand, are coils of wire that concentrate magnetic fields. Inductors will resist AC but freely pass DC. Diodes will conduct current only in one direction (see Fig. 6-10 again). Motor noise is essentially AC so it

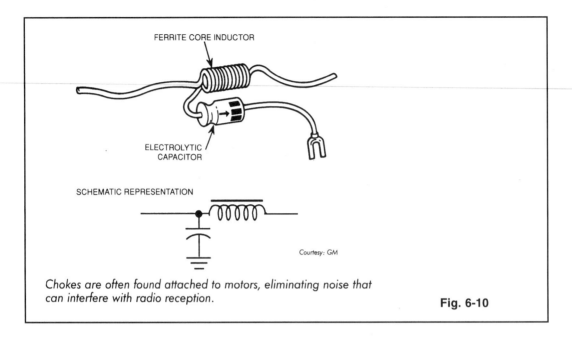

FERRITE CORE INDUCTOR

ELECTROLYTIC
CAPACITOR

SCHEMATIC REPRESENTATION

Courtesy: GM

Chokes are often found attached to motors, eliminating noise that can interfere with radio reception.

Fig. 6-10

can't easily be eliminated by a diode. Capacitors and inductors are often combined into simple circuits called "chokes" to prevent noise from interfering with a radio or cellular phone.

Suppression Diodes

Since spikes (see Fig. 6-11) from relays and solenoids always occur as DC pulses with predictable polarity, a diode can be used to conduct the voltage harmlessly to ground.

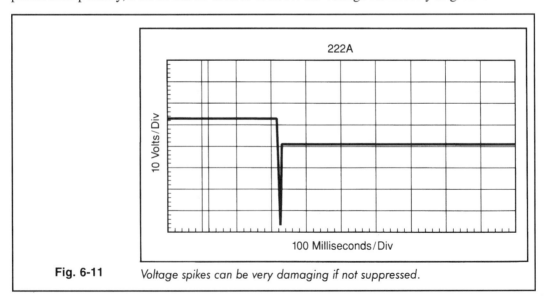

222A

10 Volts/Div

100 Milliseconds/Div

Fig. 6-11   *Voltage spikes can be very damaging if not suppressed.*

Transistor Amplifiers

Drivers are found in the computer where they power fuel injectors, relays, and solenoids. They are also found in ignition modules where they energize ignition coils. Square waves and pulse trains are produced by low-power circuits designed to shape the waveform. These circuits do not have the power to drive solenoids or even travel very far over wires. They must be amplified by circuits using power transistors called drivers. Sometimes four power transistors are combined in one integrated circuit called a quad-driver. The driver usually provides ground for the device. In this case, when looking for a waveform, connect to the ground side of the device, not the battery side. On the battery side you will see 12V DC, not the signal you want.

Whether quad-driver or individual transistors, they are mounted on heat sinks. For example, there is a driver circuit in GM's HEI (high energy ignition) module. The module uses the distributor housing as a heat sink. It must be properly mounted, using thermally conductive grease, or it will overheat.

Overheating can cause the transistor to continue to pass current even when commanded to stop—a condition called thermal runaway. High current will also overheat drivers. GM quad-drivers will limit their current flow automatically to about 1 amp, preventing thermal runaway that can destroy the transistor and possibly the device it powers.

Age and abuse, including overheating, can weaken a transistor so that it will not pass as much current as it should. In that case, the voltage on the ground side of the device being driven will not "pull down." On the scope you will see a pattern that does not fall to zero volts or as close to zero as it should. Resistance in connectors between the device and the driver can also cause this pattern (see Fig. 6-12).

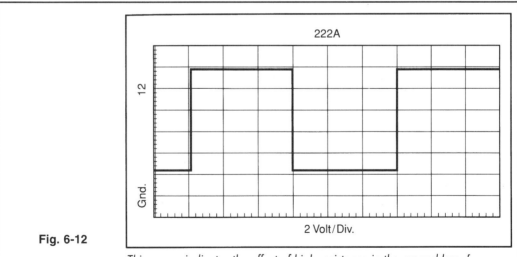

Fig. 6-12

This screen indicates the effect of high resistance in the ground leg of the circuit. The signal voltage is 12 volts that is sourced from the Ignition Module. Because the ground has high resistance, the 12 volts does not pull down low enough. The voltage drop across the corrosion is high enough that the signal is floating above ground beyond the trigger level of the module logic.

Electronic Fuel Injectors

Electronic fuel injection (EFI) is the latest step in fuel system evolution that began with the carburetor. Fuel injection allows more precise control of fuel mixture and atomizes fuel better, both resulting in more power, lower fuel consumption, and lower emission. Mechanical injection, which has been in production since the mid-1950s, is now being replaced by EFI.

There are two basic types of EFI: throttle body injection (TBI) and port fuel injection (PFI). The basic difference is the number and placement of the fuel injectors. The PFI system has one injector for each cylinder. It sprays into the intake port or manifold branch just upstream of the intake valve. In the TBI system there is one injector for each throttle bore. It sprays into the throttle assembly just upstream of the throttle plate.

How it Functions

The injector is a valve connected to the armature of a solenoid that opens it. The injection nozzle itself, called a pintle valve, is carefully ground to produce a spray pattern appropriate for the engine in which it is installed. The valve is held closed by a spring. It is opened when electric current passes through the windings surrounding the armature. Far more current is required to open the valve than to hold it open. To keep injectors cool and to reduce overall electrical consumption, many systems limit the current fed to the injector once the valve is open (see Fig. 6-13).

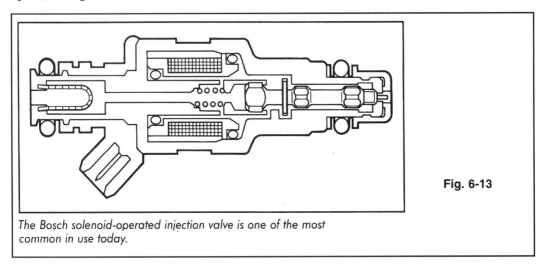

Fig. 6-13

The Bosch solenoid-operated injection valve is one of the most common in use today.

How to Hook it Up

The easiest and most reliable method of connecting the scope is to use a breakout box. If you don't have one, back-probe the connector at the ground side of the injector pin numbers. Don't forget the ground clip.

Troubleshooting

Ball valves that are less prone to clogging are now being introduced. On PFI the opening is about 0.004 inch. Pressure of about 50 PSI is necessary to force adequate fuel through the narrow passage. To open and immediately close the valve takes about 1.5 mS (0.0015 seconds) (see Fig. 6-14). Pintle valves often clog up with olefins, a wax-like substance that is a gasoline residue. Olefins bake on after the engine is shut down, particularly in hot weather, during a period known as "hot-soak."

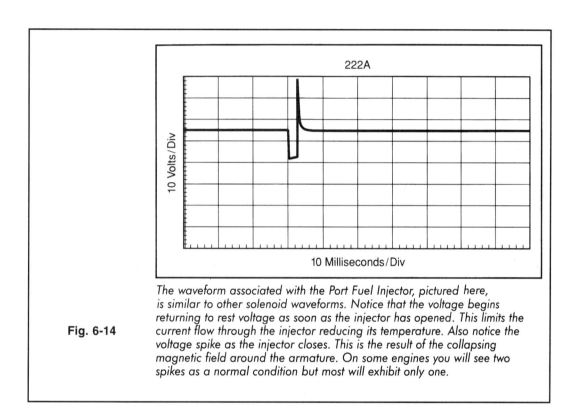

222A

10 Volts/Div

10 Milliseconds/Div

Fig. 6-14

The waveform associated with the Port Fuel Injector, pictured here, is similar to other solenoid waveforms. Notice that the voltage begins returning to rest voltage as soon as the injector has opened. This limits the current flow through the injector reducing its temperature. Also notice the voltage spike as the injector closes. This is the result of the collapsing magnetic field around the armature. On some engines you will see two spikes as a normal condition but most will exhibit only one.

Throttle Air Bypass Valve

Small engines used in today's cars produce tremendous high-speed power for their size. However, they don't have nearly the output at idle of older, larger, engines. To complicate matters, engines are now called on to drive a heavier load of accessories. Accessories switch on and off while the engine is idling, just as they do when the engine is at road speed. The throttle air bypass valve adds extra air, when the engine needs it, to maintain the specified idle speed (see Fig. 6-16).

How it Works

The idle air control motor found on many Fords and other cars is a large tapered pintle that retracts away from its seat allowing air to pass by. The pintle is attached to a solenoid that moves in response to the average current passing through its windings. The current is controlled by PWM (pulse width modulation). The greater the average pulse width, the greater the average current. And, the further the solenoid moves, the more air is bypassed (see Fig. 6-17).

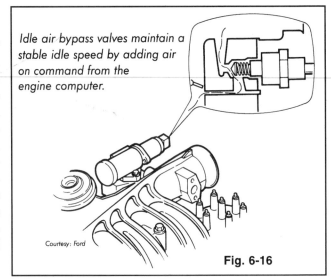

Idle air bypass valves maintain a stable idle speed by adding air on command from the engine computer.

Courtesy: Ford

Fig. 6-16

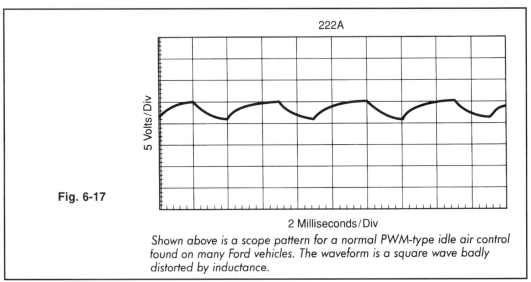

222A

5 Volts/Div

2 Milliseconds/Div

Fig. 6-17

Shown above is a scope pattern for a normal PWM-type idle air control found on many Ford vehicles. The waveform is a square wave badly distorted by inductance.

How to Hook it Up

The easiest and most reliable method of connecting the scope is to use a breakout box. If you don't have one, back-probe the connector at the ground side of the bypass valve. See the manufacturer's wiring diagram for pin numbers. Don't forget the ground clip.

Troubleshooting

If the engine does not maintain idle speed when you turn on accessories, such as the A/C compressor, there may be a problem in the Throttle Air Bypass system.

Check Scope pattern:

- There should be roughly a square wave (See Fig. 6-18).
- It should be consistent (not intermittent).
- The pulse width should vary.

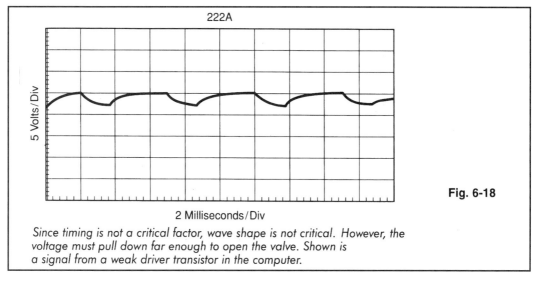

222A

5 Volts/Div

2 Milliseconds/Div

Fig. 6-18

Since timing is not a critical factor, wave shape is not critical. However, the voltage must pull down far enough to open the valve. Shown is a signal from a weak driver transistor in the computer.

If there is no square wave or it is intermittent, check the wiring and connectors for loose or corroded connectors, opens, and shorts. If the pulse width does not vary, the problem is not with the Throttle Air Bypass Valve but lies within the computer or signal inputs.

Mass Airflow Sensors

To review this vitally important component, an engine-control computer must know how much air is entering the cylinders. It needs this data to determine the exact optimal mixture of fuel with the air. One way of getting this data is from a mass airflow (MAF) sensor (see Fig. 6-19).

How it Works

All MAF sensors rely on the fact that as a resistance wire such as Nichrome gets hotter, its resistance increases. This is called negative temperature coefficient (NTC). In an MAF sensor, a Nichrome wire is stretched across the throat of the air intake where the air rushing past will cool it. The computer supplies a constant source of current to the wire, and a voltage-to-frequency converter circuit inside the MAF converts the voltage drop across the wire into a square wave. The frequency of the square wave changes depending on the airflow (see Fig. 6-20).

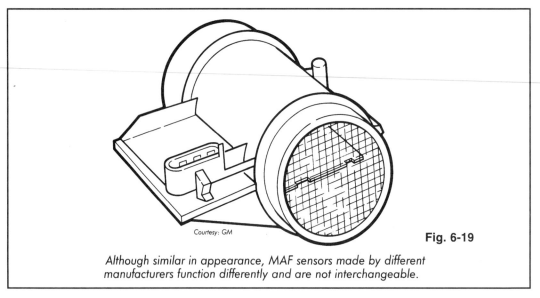

Courtesy: GM

Fig. 6-19

Although similar in appearance, MAF sensors made by different manufacturers function differently and are not interchangeable.

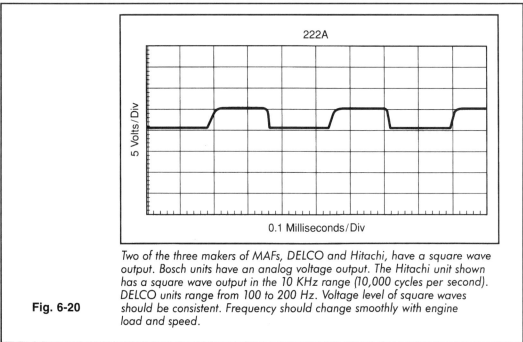

Fig. 6-20

Two of the three makers of MAFs, DELCO and Hitachi, have a square wave output. Bosch units have an analog voltage output. The Hitachi unit shown has a square wave output in the 10 KHz range (10,000 cycles per second). DELCO units range from 100 to 200 Hz. Voltage level of square waves should be consistent. Frequency should change smoothly with engine load and speed.

How to Hook it Up

The easiest and most reliable method of connecting the scope is to use a breakout box. If you don't have one, back-probe the connector at the MAF sensor. See the manufacturer's wiring diagram for pin numbers. Don't forget the ground clip.

Troubleshooting

Watch the scope pattern while an assistant drives the car or while on a road simulator. The pattern should not have gaps (see Fig. 6-21). Sometimes tapping on the sensor with a pencil or small screwdriver will cause an intermittent fault to show up. The unit must be replaced if it has intermittent faults.

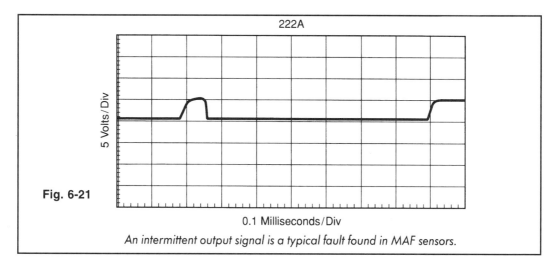

Fig. 6-21

222A

5 Volts/Div

0.1 Milliseconds/Div

An intermittent output signal is a typical fault found in MAF sensors.

Delco CS Series Generators

How it Functions

Delco CS series generators have been in production since 1986 (see Fig. 6-22). They are not only visibly different form the older SI series but function differently, too. The CS series has an internal digital regulator, controlling field current by pulse width modulation , whereas the older SI series used an analog regulator. The CS series has no diode trio, getting power instead from the main diodes. The CS series will light the charge indicator if the output voltage goes either too high or too low. The regulator is designed to change output slowly, delaying full changes by 2.5 seconds. The delay allows the ECM to keep the idle speed stable.

There are two basic CS generators supplied by Delco—PLIS and PLFS. They are known by the terminals on the regulator. There are also rebuilt generators available that have a variable frequency regulator and may have either or both designations. They may not be computer compatible.

CS Series Terminal Designations

- P Terminal—Connected to one stator winding, providing an AC signal for a tachometer or other device. It does not need to be connected for the generator to work.

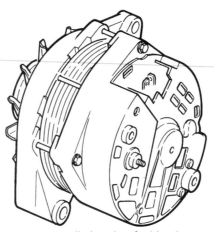

Fig. 6-22

CS series generators can usually be identified by the terminal designations at the rear of the case, PLIS or PLFS. Check the signal at the F terminal with your scope. You should be able to monitor the generator's response to changing loads. Probe should be attached to the F terminal.

- L Terminal—This receives current from the body control module (BCM) and signals that the regulator is operating. This happens only on BCM-equipped cars. L terminal lights the indicator lamp and powers the regulator on non-BCM cars.
- I Terminal—This receives voltage from the ignition, sometimes through a resistor.
- F Terminal—This provides a signal to the body control module (BCM) telling it how hard the generator is working. It is not always connected.
- S Terminal—This senses battery voltage.

How to Tell Them Apart

If you install a PLFS unit in a car intended to have a PLIS, the regulator may be damaged. Application charts may not specify which unit each car requires. Before you install a new unit, measure the voltage at the harness connector I or F terminal with the key on. If you measure 12 V at this terminal the car must have a PLIS unit installed (see Fig. 6-23).

To determine which type generator you need to install, look at the regulator's electrical connector. It may be marked with either PLIS or PLFS, indicating which type of regulator can be found in the generator. However, the connector may not be marked, or marked in a confusing manner—such being marked with both configurations. Remanufactured generators may not have the correct regulator regardless of how they are marked. To be sure, connect only the main current output terminal for the alternator, known as BAT, and the L terminals, then follow the steps in the troubleshooting paragraph *before* you connect the wiring harness plug.

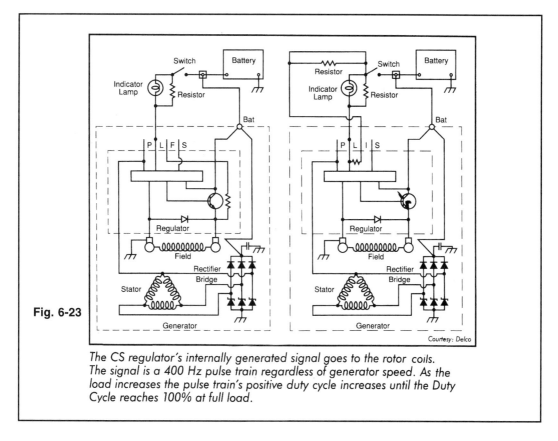

Fig. 6-23

Courtesy: Delco

The CS regulator's internally generated signal goes to the rotor coils. The signal is a 400 Hz pulse train regardless of generator speed. As the load increases the pulse train's positive duty cycle increases until the Duty Cycle reaches 100% at full load.

How to Hook It Up

To monitor the regulator's PWM pattern, connect the scope to the F terminal at the back of the generator. Depending on the generator's position, you may need to use a jumper wire. To monitor the generator's output, connect the scope to the BAT terminal. Don't forget to connect the ground clip.

You should see a pulse train with fixed frequency and variable duty cycle. On PLFS units probe the F terminal. On PLIS units, probe the positive brush holder (see Fig. 6-22 again). If the frequency changes, the regulator is an after-market unit that may not be compatible with the body control module.

Troubleshooting

CAUTION: Do not attempt to full-field a CS series generator. The regulator is internal and there is no test hole. Do not apply battery voltage to the L or F terminals.

Before you make any electrical tests make the usual checks of wiring, belts, brackets, etc. Next, determine whether the unit is a PLIS or PLFS by following steps 1 through 5. The P, L, and S terminals will always be in the same positions (see Fig. 6-22 again), leaving one terminal unknown. It will be either I or F. Your scope and voltage tests will tell you whether the unit is good.

1. Check that there is voltage to the regulator L terminal; if there is no voltage, correct the problem in the BCM or wiring.

2. Connect a carbon pile across the battery to provide a load for the generator. Make sure all accessories are off.

3. Connect the scope to the unknown terminal.

4. Run the engine at about 3,000 RPM.

5. Watch the pattern on the scope while varying the load with the carbon pile between 10 amps and the generator's rated output.

   Conclusion: The unit is good if the duty cycle of the pulse train varies with load and the unit generates 85 percent of its rated output or more.

   To identify PLIS or PLFS, as you perform steps 1 through 5 just cited, notice if your scope pattern changes from a train of narrow pulses to a square wave and finally becomes straight-line DC at battery voltage.

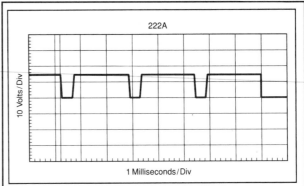

CS-144 Alternator — At left is terminal "F" waveform when the key is on but the engine is off. Note that the frequency is about 400 Hz and the duty cycle is about 30%. Duty cycle is limited when generator is not turning. At right is terminal "F" waveform after the engine has started. At right, note the higher but decreasing duty cycle as the battery recovers from the starter current draw. **Fig. 6-24**

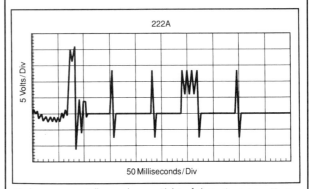

Modern cars, especially upscale cars with lots of electronics, are sensitive to voltage spikes and AC voltage resulting from bad diodes. If there is a lot of "hash" (noise) coming out of the generator, repair or replace the generator before it ruins sensitive electronic components. **Fig. 6-25**

If it does, the unit is a working PLFS. If there is no pulse train, the unit is either a PLIS or a nonworking PLFS. Check the voltage output with the unit under full load to determine whether it works or not. Figures 6-24 and 6-25 show, first, a CS-144 alternator's correct waveform and, second, both voltage surges and spikes occurring together. These are potentially damaging alone, but are a real killer when they are present together!

Variable Reluctance Sensor

How It Functions

This is the most common type of magnetic sensor (see Fig. 6-26). It generates its own voltage and is used as a crankshaft sensor or wheel-speed sensor. The sensor consists of a soft iron core inside a magnet with a coil of fine wire around it. The sensor is mounted close to a toothed iron wheel. As a tooth moves toward the sensor's core, the magnetic field surrounding the wire coil gets stronger. As the tooth moves away, the field weakens. The changing field induces an AC voltage at the terminals of the sensor.

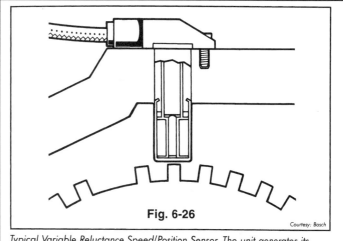

Fig. 6-26

Courtesy: Bosch

Typical Variable Reluctance Speed/Position Sensor. The unit generates its own pulses as the teeth move past the iron core.

How to Hook It Up

The easiest and most reliable method of connecting the scope is to use a breakout box; however, on some systems the VR sensor signal will not be a available at the BOB. If that is the case, or if you don't have a BOB, back-probe the VR sensor connector with the probe at one terminal and the ground clip at the other (see Fig. 6-27). See the manufacturer's wiring diagram for pin numbers. Don't forget the ground clip.

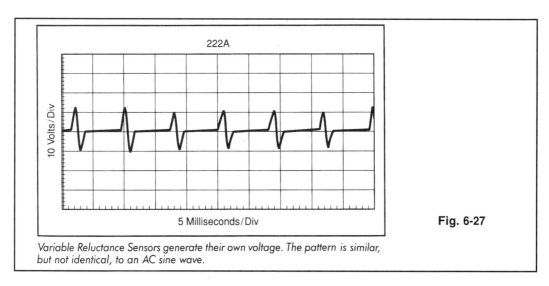

Fig. 6-27

Variable Reluctance Sensors generate their own voltage. The pattern is similar, but not identical, to an AC sine wave.

Troubleshooting

- No output—Test sensor resistance and replace if open or shorted.
- Low output—Reduce air gap. Replace sensor if output is still low.
- High output—Increase air gap.
- Variable output—(See Fig. 6-28) Replace or straighten crankshaft or toothed wheel.

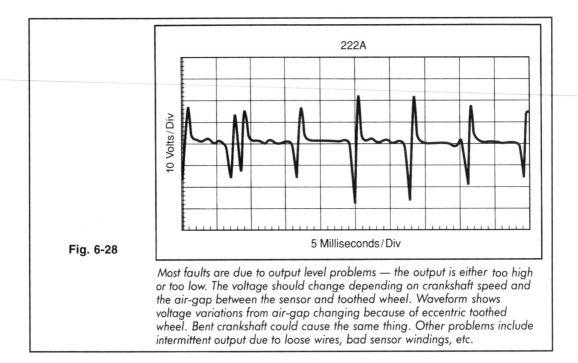

222A

10 Volts/Div

5 Milliseconds/Div

Fig. 6-28

Most faults are due to output level problems — the output is either too high or too low. The voltage should change depending on crankshaft speed and the air-gap between the sensor and toothed wheel. Waveform shows voltage variations from air-gap changing because of eccentric toothed wheel. Bent crankshaft could cause the same thing. Other problems include intermittent output due to loose wires, bad sensor windings, etc.

Vehicle Speed Sensor

The vehicle speed sensor provides vehicle speed information to the computer, the cruise control, and the speedometer. Among other things, the computer uses the data to decide when to engage the torque converter clutch.

There are two types—photo-electric (PE) and permanent magnet (PM) (see Fig. 6-29). The PM type is usually mounted directly on the transmission or transaxle, while the PE type is driven by a conventional cable and is found under the dash.

How It Functions

The PM type is simply a permanent magnet AC generator. Its output is an AC sine wave. Its voltage output, like a variable reluctance sensor, is dependent on speed. At 2 MPH the voltage output is 5V peak to peak and at 60 MPH it is 40V peak.

The VSS output signal is fed to the ECM or a buffer/amplifier mounted under the dash. The buffer converts the sine wave to a square wave and feeds it to the speedometer, computer, and cruise control. At the output of the buffer you should see a 5V square wave (see Fig. 6-30).

The photo-electric (PE) type sensor works very differently. It is typically mounted under the dash and driven by a cable. Attached to the cable is a propeller-like device that is mirror-finished on one side. The light from an infrared light emitting diode (LED) is reflected off the

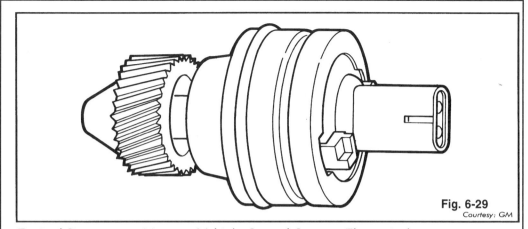

Fig. 6-29

Courtesy: GM

Typical Permanent Magnet Vehicle Speed Sensor. The unit shown mounts directly to the transaxle. The wiring harness connects to terminals at the right.

Fig. 6-30

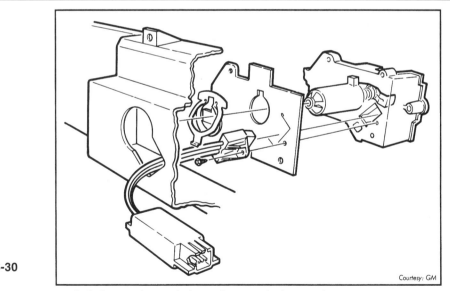

Courtesy: GM

Exploded view of typical Photo Electric Vehicle Speed Sensor installation. It is mounted in the instrument cluster. Its output waveform will be a square wave.

propeller blades and picked up by a photo cell. The infrared light is not visible to the eye. The signal from the photo cell triggers a transistor in the buffer circuit, which converts it to a square wave usable by the computer.

How to Hook It Up

The easiest and most reliable method of connecting the scope is to use a breakout box; however, on some systems the VSS sensor signal will not be available at the BOB. If that is the case, or if you don't have one, back-probe the VSS sensor connector with the probe at one terminal and the ground clip at the other. See the manufacturer's wiring diagram for pin numbers. Don't forget the ground clip.

Troubleshooting

When troubleshooting a missing VSS signal, remember to check the fuse first. If there is no power to the buffer, there will be no square-wave output. If the fuse is good, check the easy things first. The PM generator, for example, may be easier to get to than a buffer mounted under the dash. With the photo-electric type, remove the instrument cluster from underneath the dash.

Note: If you have a sine wave coming from the PM generator, but no square wave coming from the buffer, don't assume the problem is in the buffer. Check the signal coming into the buffer; it may not be there because of a loose connector between the PM generator and the buffer (see Fig. 6-31, a VSS scope pattern from a car driven at 42 mph).

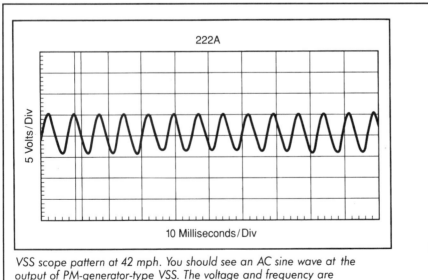

Fig. 6-31

VSS scope pattern at 42 mph. You should see an AC sine wave at the output of PM-generator-type VSS. The voltage and frequency are dependent on speed.

Revisiting Suppression Diodes

In electromechanical devices such as solenoids, electric motors, relays, or fuel injectors, energy is stored in the electromagnetic field surrounding the device. For example, the magnetic field is maintained and is stationary as long as steady current flows through the windings of an air-conditioning compressor clutch. The field collapses and the energy is released when current to the clutch is shut off (see Fig. 6-32).

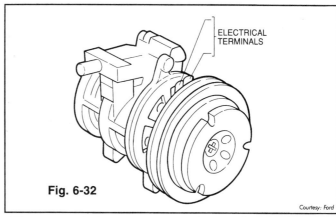

ELECTRICAL TERMINALS

Fig. 6-32

Courtesy: Ford

A/C compressor clutches are among many electro-mechanical components that can cause problems if suppression diodes fail.

The stored magnetic energy is transformed into electric current when the now-moving (collapsing) field cuts across the windings. The effect of this current travels back to the control circuit as a voltage spike.

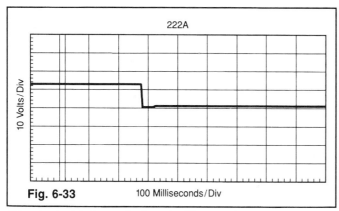

222A

10 Volts/Div

Fig. 6-33   100 Milliseconds/Div

When the clutch is de-energized you should see simply a transition from high to low voltage. This waveform shows a clutch with a good diode. Note the slightly negative voltage after the transition, it represents the forward turn-on bias voltage of diode.

The spike can reach minus 100V or more and can damage electronic components. Since the spikes always occur as predictable voltage, a diode can be used to conduct the voltage harmlessly to ground (see Fig. 6-33). Blocking diodes are sometimes also placed in series with a control circuit or switch. Look at a wiring diagram for any circuit controlled by relays and you will see at least one diode for each relay (for example, power-window motors).

Troubleshooting

Connecting the scope to find a voltage spike is a little different in that you may see the voltage spike anywhere in the electrical system, including ground. You might start by connecting the probe to the B+ contact in the cigarette lighter and operate various accessories until you see a spike. Then connect to the component you suspect. As always, don't forget the ground clip.

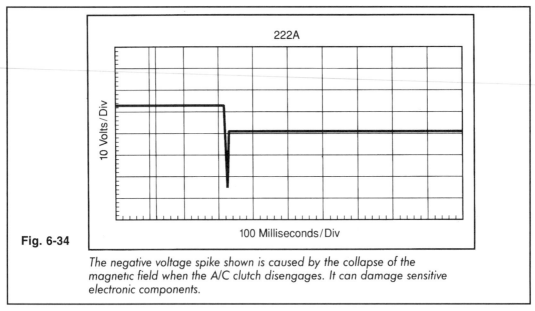

Fig. 6-34

222A

10 Volts/Div

100 Milliseconds/Div

The negative voltage spike shown is caused by the collapse of the magnetic field when the A/C clutch disengages. It can damage sensitive electronic components.

Troubleshooting consists mainly of recognizing that that there are faulty diodes and then finding them. You may hear a "pop" in the AM radio when the offending component switches off. A more serious symptom is repeated replacement of the computer. Spikes sometimes occur on the computer's ground so check here with the scope while operating different electro-mechanical devices (see Fig. 6-34). They may also be present at the cigarette lighter socket although the loading effect of the wiring may diminish them somewhat. Check at the cigarette lighter while operating different devices. The list of devices that use diodes is too long to include here and varies from car to car; familiarize your self with the wiring diagram of the car you're working on.

Hall Switch

Incidentally, if this subject interests you, there is a bibliography and a list of suppliers at the end of this chapter.

How It Functions

As review, you may recall that a Hall sensor, or Hall switch as it is sometimes called, is most often used as a crankshaft sensor or camshaft sensor (see Fig. 6-35). It tells the computer the engine's crankshaft speed and, on some systems, crank position. On GM combination/dual sensors it provides speed and No. 1 cylinder TDC (top dead center) data. On some others it provides continuous crank-position data.

Inside the sensor is a Hall-effect cell that responds to a changing magnetic field by changing the amount of current it will conduct. An interrupter wheel (vane wheel), with slots cut in it, rotates between the Hall cell and a magnet, alternately exposing it to the magnetic field. The cell conducts more as the slots in the vane wheel pass. The signal produced is amplified by a transistor mounted in the sensor housing and sent to the computer. The sensor must have external power. Unlike the variable reluctor-type sensor, the Hall sensor does not produce its own voltage.

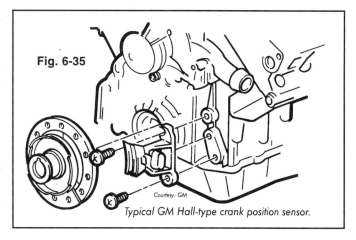

Fig. 6-35

Courtesy: GM

Typical GM Hall-type crank position sensor.

How to Hook It Up

The best method of connecting the scope is to use a breakout box. If you don't have one, back-probe the connector at the ignition module. See the manufacturer's wiring diagram for pin numbers. Don't forget the ground clip.

Troubleshooting

If the maximum voltage is too low, check for resistance in the signal wire and connectors between the computer and sensor. Make the same check of the sensor's power wire. If you find resistance (manufacturers specify maximum resistance in their manuals), clean all associated connector terminals. If resistance is normal, suspect the sensor. A weak magnet can cause low output. Some sensors use a separate magnet that can be replaced but most have an integral magnet requiring sensor replacement should it fail.

If the minimum voltage is too high, check for resistance between the sensor and ground. If you find resistance (it should be nearly zero Ohms), clean all the contacts between the sensor and ground. If there is no resistance, suspect a faulty sensor (see Fig. 6-36).

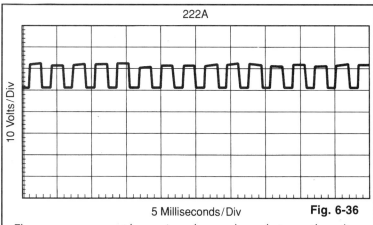

222A

10 Volts/Div

5 Milliseconds/Div   **Fig. 6-36**

The square wave must have crisp edges and vary between the voltages specified. The transition from low to high or high to low is important. The manufacturer specifies the voltage levels required for the computer to recognize an "on" or "off" state.

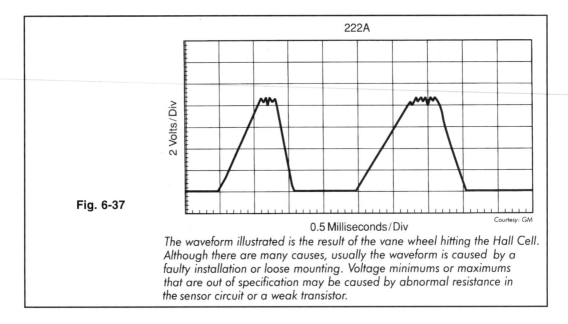

Fig. 6-37

222A

2 Volts/Div

0.5 Milliseconds/Div

Courtesy: GM

The waveform illustrated is the result of the vane wheel hitting the Hall Cell. Although there are many causes, usually the waveform is caused by a faulty installation or loose mounting. Voltage minimums or maximums that are out of specification may be caused by abnormal resistance in the sensor circuit or a weak transistor.

If the waveform indicates mechanical interference, check for damaged vanes on the vane wheel, see Fig. 6-37. Also check and adjust the sensor for proper clearance.

Throttle Position Sensor (TPS)

How It Functions

The throttle position sensor tells the computer the following things:

- Throttle opening.
- Whether, and how fast the throttle is opening.
- Whether, and how fast the throttle is closing.
- When the throttle is wide open.
- When the throttle is at idle.

One of its most important functions is to tell the computer that the throttle is opening. It replaces the accelerator pump found on carbureted engines, stopping the engine from stumbling when the throttle is opened quickly. When that happens, manifold pressure (MAP) quickly rises (vacuum drops)

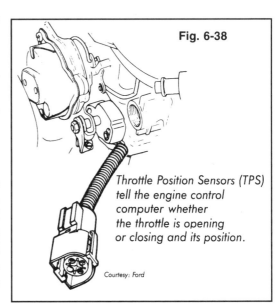

Fig. 6-38

Throttle Position Sensors (TPS) tell the engine control computer whether the throttle is opening or closing and its position.

Courtesy: Ford

causing vaporized gasoline to condense on the manifold walls. Since there is less fuel available to the cylinders, more must be added to the air stream.

Another important function is to tell the computer that the throttle is closing (see Fig. 6-38). To maintain acceptable emissions, the computer must lean out the mixture when MAP drops (vacuum rises) (see Fig. 6-39). For the best fuel economy, the computer completely shuts off fuel in some engines when vacuum is high and the throttle is at idle. Therefore, the computer must know when the throttle is at idle.

Throttle position information is a variable DC voltage from a potentiometer (variable resistor) attached to the throttle valve. WOT (wide open throttle) and throttle-closed signals come from switches attached to the TPS.

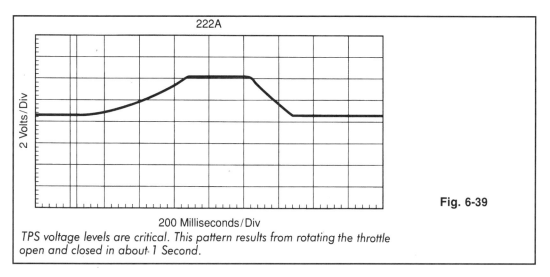

222A

2 Volts/Div

200 Milliseconds/Div

Fig. 6-39

TPS voltage levels are critical. This pattern results from rotating the throttle open and closed in about 1 Second.

How to Hook It Up

The best method of connecting the scope is to use a breakout box. If you don't have one, back-probe the connector at the signal return side of the TPS. See the manufacturer's wiring diagram for pin numbers. Don't forget the ground clip.

Troubleshooting

Throttle position sensors usually aren't repairable. Although they sometimes can

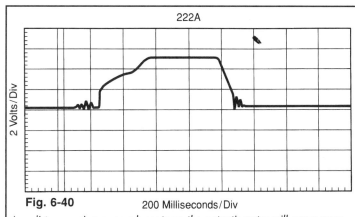

222A

2 Volts/Div

Fig. 6-40   200 Milliseconds/Div

Any dirt, corrosion or rough spots on the potentiometer will cause poor drivability. Watching the voltage change with a scope will detect faults that cause drivability problems long before a voltmeter will detect the problem.

be cleaned with electrical contact cleaner, the best policy is to replace them if they show a faulty scope pattern. A dirty or damaged TPS can cause a great many problems when you drive it and often will not let an error code even if a legitimate fault definitely exists. It's best to include this among the first items to check when hunting down such troubles (see Fig. 6-40).

Ford TFI

How It Functions

The most common ignition system found on Ford vehicles has been dubbed TFI for thick film ignition (see Fig. 6-41). This system uses a Hall switch in the TFI module, mounted on the distributor, to produce a basic timing signal PIP (profile ignition pickup). This signal is sent to the EEC-IV module (engine control computer). The EEC-IV module computes ignition timing and sends a signal, SPOUT (Spark Out), back to the TFI module, which then fires the coil.

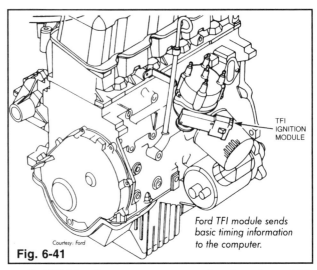

TFI IGNITION MODULE

Courtesy: Ford

Ford TFI module sends basic timing information to the computer.

Fig. 6-41

PIP is a symmetrical square wave. SPOUT is also a square wave, but is pulse-width modulated. Its rising and falling edges move in relation to PIP. The rising edge controls spark timing and the falling edge controls coil saturation (dwell).

Watching both simultaneously on the Tek 222A Dual-Trace Scope will tell you whether the EEC-IV can compute timing based on sensor inputs. For example: If the BP/MAP sensor fails, the rising edges of SPOUT will not move relative to the rising edges of PIP when manifold absolute pressure changes.

How to Hook It Up

Again, the best method of connecting the scope is to use a breakout box. If you don't have one, back-probe the connector at the TFI module. See the manufacturer's wiring diagram for pin numbers.

Troubleshooting

Changing engine speed or manifold vacuum should have the greatest effect on ignition timing. If changing vacuum has no effect on the rising edges of SPOUT, look for a faulty BP/MAP.

If PIP is absent, the engine will not start; look for a bad TFI module or other distributor problem.

IF SPOUT is absent, the system is probably in LOS (limited operation strategy) or limp-home mode. Look for problems in the EEC-IV module or bad wiring harness connectors.

If the leading edges of PIP or SPOUT are rounded, timing will be inaccurate. Look for problems in the module producing each signal (see Fig. 6-42).

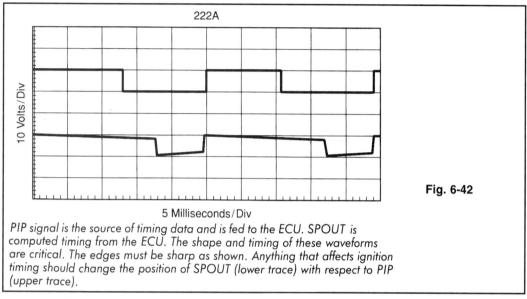

222A

10 Volts/Div

5 Milliseconds/Div

Fig. 6-42

PIP signal is the source of timing data and is fed to the ECU. SPOUT is computed timing from the ECU. The shape and timing of these waveforms are critical. The edges must be sharp as shown. Anything that affects ignition timing should change the position of SPOUT (lower trace) with respect to PIP (upper trace).

Caution: Make sure that all scope leads and cords are out of the way of the fan, belts, and pulleys each time the engine is to be started.

Caution: Make sure that scope leads are also away from ignition wire. They generate a magnetic field that it strong enough to be picked up by the scope giving an unexpected pattern (see Fig. 6-43).

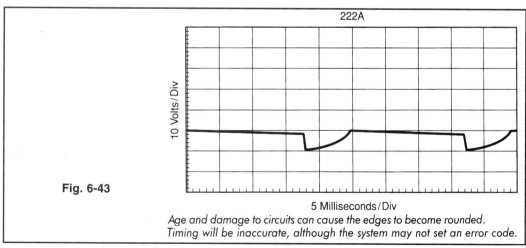

222A

10 Volts/Div

Fig. 6-43

5 Milliseconds/Div

Age and damage to circuits can cause the edges to become rounded. Timing will be inaccurate, although the system may not set an error code.

Caution: DO NOT, under any circumstances, connect scope probes to ignition secondary voltages. It will damage the scope.

Caution: Review ELECTROSTATIC DISCHARGE (ESD) precautions before connecting any test equipment. Do not remove any control module from the vehicle without observing ESD precautions. ESD precautions include using a grounding clip to your wrist that has an alligator clip test pinching metal somewhere on your car.

Caution: ALWAYS connect the scope's ground clip to a ground reference point for the circuit on which you are working. Failure to do so will cause faulty scope patterns. Extend the scope's ground wire only if absolutely necessary. Excessive ground-lead length will distortion scope patterns.

The Tektronix ScopeMeter™

This instrument, the TekMeter THM565 (see Fig. 6-1 again) is so small its size resembles a paperback novel—yet it is quite powerful! This handheld instrument is both a true rms DMM and a DSO that can handle signals to 5 MHz. This is more than adequate for any car troubleshooting you will ever encounter, unless you attempt to troubleshoot a car radio. This instrument's front panel has a dozen ruggedized controls under and to the right of its small, but bright, 2.5" x 4.8" LCD. As with any scope, you place the leads into the proper terminals (see Fig. 6-44). If you attempt to test a heavy current drain and in-rush, such as when you crank your car, you'll need a clamp-on current clamp (see Fig. 6-45). Figures 6-46 and 6-47 show how to make power measurements. Note that just like the formula for power, which is: Power (Watts) = Voltage (Volts) x Current (Amps), Fig. 6-46 shows how CH1 (Channel 1) is hooked to measure current by use of a current transformer. Channel 2 is hooked up across the unit under test and measures its voltage. The TekMeter combines these by multiplying them to give you an indirect power reading. That is, Fig. 6-47 shows both current and voltage displayed. To

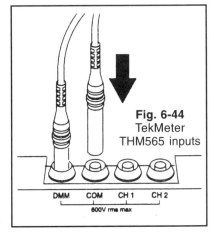

Fig. 6-44
TekMeter
THM565 inputs

derive a power reading, you would need to manually multiply both quantities with maybe a pencil and paper, or a calculator.

To observe the line waveform, you would first push the METER/SCOPE button and switch the hot input (red) lead from the DMM to the CH1 connector. Don't move the common lead (black) since it is at ground already. After a few seconds, autoranging settles down and you have a display that is quite read-

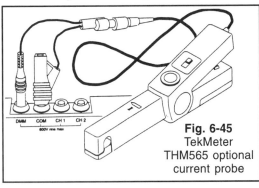

Fig. 6-45
TekMeter
THM565 optional
current probe

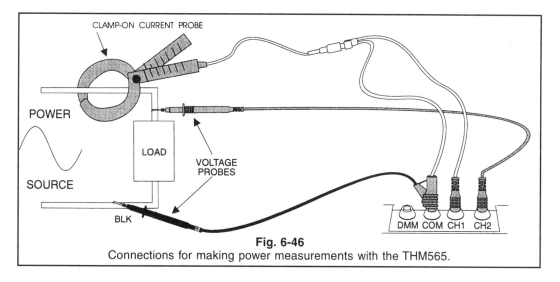

Fig. 6-46
Connections for making power measurements with the THM565.

able. If you wanted to make a power measurement with this instrument you would need to first realize that power is the product of voltage times current. That is, power in Watts = voltage in (Volts) multiplied by current in (Amps). Since you have a DMM here, you can use it to measure current and then the scope can measure voltage (see Fig. 6-46 again). This resulting combined reading is voltage times current or power, expressed in Watts. In fact, if you look closely at Fig, 6-47, you will see both voltage bad current waveforms.

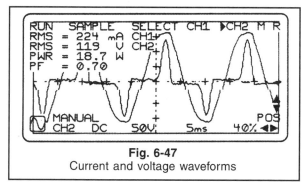

Fig. 6-47
Current and voltage waveforms

As you probably recall, the convention called PWM often controls an electrical automotive device by varying its duty cycle (ratio of on-time versus off-time). You can also view this on this DSO (see Fig. 6-48). Thus far, we have encountered numerous PWM (pulse width modulation) schemes, most of which controlled the amount of time an electronic device was on (its duty cycle). It accomplishes this by controlling, for example, a transistor that may have held a relay open. That also directly relates to the power it allows to be dispensed or controlled.

Note in Fig. 6-48(a) the upper waveform on the TekMeter's screen. Both waveforms are from control circuits that use a form of PWM. However, the one on the top is powered by AC, such as you would find in your home's light dimmer switches. The fact that it is powered by AC is apparent in the shape of the waveforms. They look like, and indeed they are, AC waveforms that have been clipped or cut off. You will rarely, if ever, observe this anywhere in your car's electrical system.

Contrast this to the lower waveform in Fig. 6-48(b). This also shows a PWM approach taken to control a power device. However, here there are sharp, crisp, well-defined square waves with fast rise times. This sudden turning on and turning off is obviously what makes a square wave just exactly that. This indicates that these are being supplied power in the form of DC, probably 12 VDC from your car's battery.

If you would wish to monitor the output of an IGBT (Insulated Gate Bipolar) or ordinary non-FET-type power transistor, you'd monitor its collector and hook onto (use as a reference to ground) its emitter, which is at ground level. You may easily accomplish this with

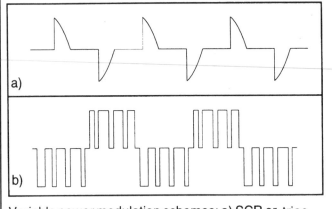

Variable power modulation schemes: a) SCR or triac duty cycle modulation; b) digital PWM power modulation. **Fig. 6-48**

the TekMeter (see Fig. 6-49). This shows the output you would expect from an IGBT-type transistor.

Note again the presence of sharp, crisp, very distinct and well-defined in shape and form waveforms that are ideal square waves. This again is the telltale sign that this IGBT transistor is being powered by DC. As a note, rarely will you see a transistor powered by AC, and when you do, it is only on its base—never its collector!

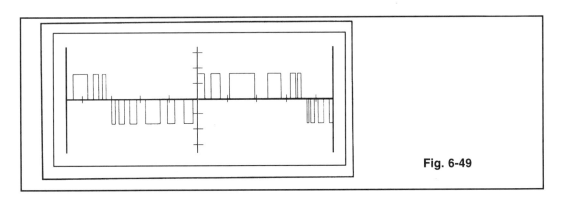

Fig. 6-49

The Fluke 97 and Auto Troubleshooting

We have spent considerable time explaining both the theory of operation, and the actual controls and indicators that are easily at your disposal with this powerful instrument. Therefore, let's examine some actual applications. The following heavily illustrated section will rely greatly upon the DSO screen patterns you should observe in various situations while troubleshooting your car's electrical system.

In the following screen shots you will see a lot of data that pertains to such vital factors as the time base selected and the voltage scale, plus if the scope probe is a 1:1, 10:1, or even a 100:1 or above. Starting with these probes, see Fig. 6-50, you will see what they look like as well as where you plug them into this instrument's back panel. Figure 6-51 shows the selection you have over using a 1:1 or 10:1 probe. One unique and very handy feature of the model 97 is that it has a built-in function generator capable of producing sine waves, square waves, and several other type of waveforms at variable frequencies and voltage amplitudes. Figure 6-52 shows the generator feature used—note the blackened in area where the word GENERATE appears. This is in contrast to the measure and print modes.

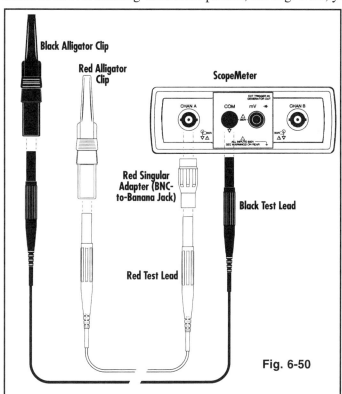

Fig. 6-50

Figure 6-53 shows what a typical display looks like. Here you see that it is in the mV (millivolt range) and also note the 049.8 D%, which means a 49.8 percent duty cycle square wave (almost symmetrical.) The other 976.6 Hz is the frequency of the signal. As you view these screens, items highlighted will be very intuitively obvious to you and these will soon become almost second nature to you. If you want to really get fancy, there is an optical interface as

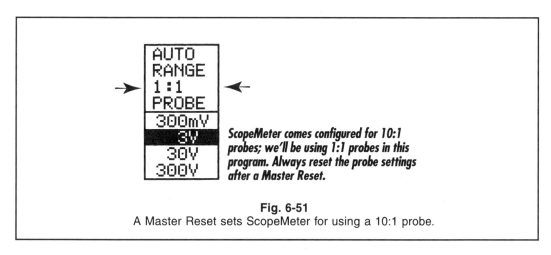

Fig. 6-51
A Master Reset sets ScopeMeter for using a 10:1 probe.

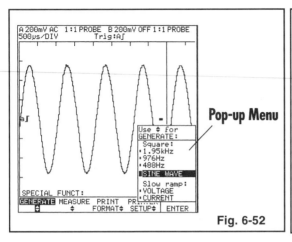

Pop-up Menu

Fig. 6-52

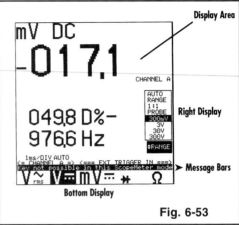

Display Area

Right Display

Message Bars

Bottom Display

Fig. 6-53

well since there now exists many optical sensors within the more modern cars. In the following material you will see many references made to you car's all-important sensors, which are growing in number, complexity, and levels of sophistication monthly. However, when referring to a sensor when speaking to a garage mechanic, call it a sending unit, which is the name they almost exclusively use to describe it.

Troubleshooting Faults with the ScopeMeter

Let's start simply by observing an AC signal, see Fig. 6-54. This is from a Ford ABS wheel sensor, and as your car's wheels rotate at a constant speed, you can expect a relatively undistorted sine wave. Another AC application is when you measure ripple on your alternator. Ripple is a small amount of AC that rides on top of DC (see Fig. 6-55). To capture this signal in Fig. 6-55 you may have to use a low-pass filter. This is a resistor connected to a capacitor. You apply power to the resistor and ground the capacitor. You take your measurement at the intersection of these two components. The capacitor passes the higher frequency AC to ground, but blocks the DC. This helps combat the effects of electrically generated noise, which can attain an amplitude of 500 mV or ½ Volt.

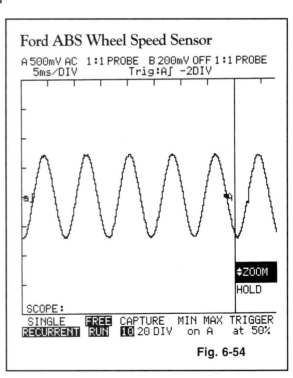

Ford ABS Wheel Speed Sensor

Fig. 6-54

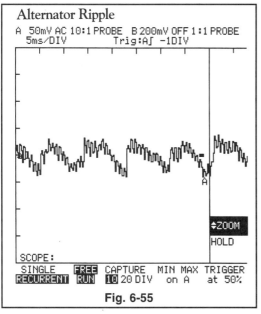

Fig. 6-55

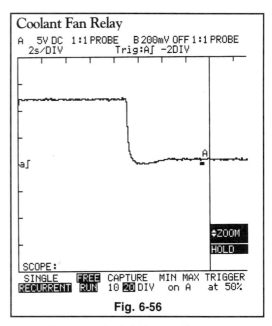

Fig. 6-56

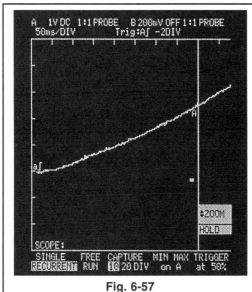

Fig. 6-57

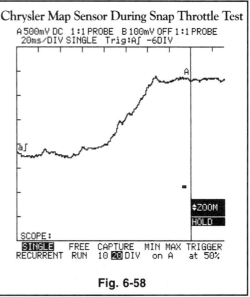

Fig. 6-58

Figure 6-56 shows how your car's computer can command a transistor-based switch, or similar active device, to engage an actuator, motor, relay, solenoid, etc. (something that engages, catches, moves another electromechanical device). By definition, an active device is an electronic component composed of transistors so that it performs some form of amplification or other function on a signal applied to it. In contrast, a passive device is one of the following three—a resistor, capacitor, or inductor.

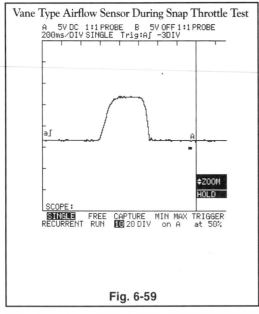

Vane Type Airflow Sensor During Snap Throttle Test

A 5V DC 1:1 PROBE B 5V OFF 1:1 PROBE
200ms/DIV SINGLE Trig:A∫ -3DIV

a∫

A

♦ZOOM
HOLD

SCOPE:
SINGLE FREE CAPTURE MIN MAX TRIGGER
RECURRENT RUN **10** 20 DIV on A at 50%

Fig. 6-59

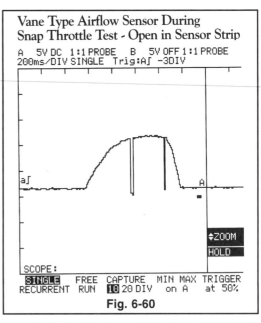

Vane Type Airflow Sensor During Snap Throttle Test - Open in Sensor Strip

A 5V DC 1:1 PROBE B 5V OFF 1:1 PROBE
200ms/DIV SINGLE Trig:A∫ -3DIV

a∫

A

♦ZOOM
HOLD

SCOPE:
SINGLE FREE CAPTURE MIN MAX TRIGGER
RECURRENT RUN **10** 20 DIV on A at 50%

Fig. 6-60

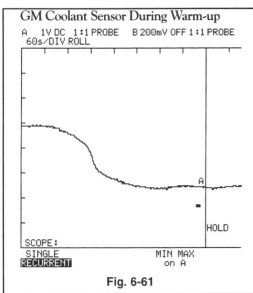

GM Coolant Sensor During Warm-up

A 1V DC 1:1 PROBE B 200mV OFF 1:1 PROBE
60s/DIV ROLL

A

HOLD

SCOPE:
SINGLE MIN MAX
RECURRENT on A

Fig. 6-61

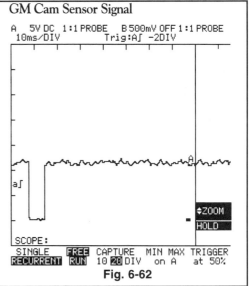

GM Cam Sensor Signal

A 5V DC 1:1 PROBE B 500mV OFF 1:1 PROBE
10ms/DIV Trig:A∫ -2DIV

a∫

A

♦ZOOM
HOLD

SCOPE:
SINGLE **FREE** CAPTURE MIN MAX TRIGGER
RECURRENT **RUN** 10 **20** DIV on A at 50%

Fig. 6-62

Figure 6-57 shows the reverse video feature of this DSO. That is, where there was formerly black now appears as white and vice versa. This waveform is of a ramping-up waveform; therefore, if spikes exist here, you have a short to the power source (usually your car's 12 VDC battery.) This illustration shows how your TPS looks upon being opened. Conversely, on a ramping-down waveform you have a short to ground.

Figure 6-58 shows a rather abrupt transition of the MAP sensor when you press the accelerator on a Chrysler product. Contrast this to Fig. 6-59 in which you have a vane-type air sensor with smooth quiet (and very acceptable) transitions from a high to low signal, and then

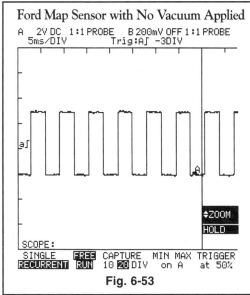

Fig. 6-53

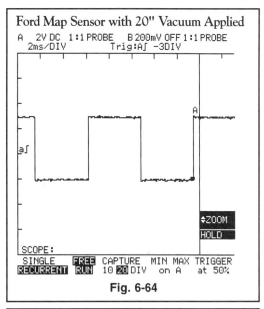

Fig. 6-64

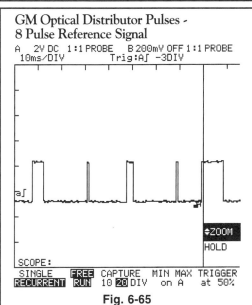

Fig. 6-65

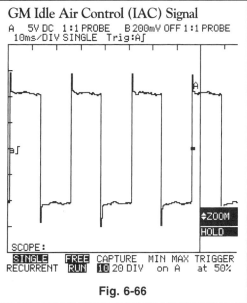

Fig. 6-66

recycling back again. Figure 6-60 shows this same vane-type air sensor with glitches or voltage spikes. There are two possible causes for this unaccceptable output voltage waveform. One is that there is a short to ground and the other reason is an electrical open existing in the resistive strip somewhere over its operating range.

Returning to troubleshooting our cooling system, Fig. 6-61 is a screen that illustrates how a thermistor with an NTC (negative temperature coefficient) has its resistance decrease as temperature increases, and vice versa.

Figure 6-62 shows how your car's cam sensor should appear when working properly. As a possible review and background, you recall that these types of sensors determine either your crankshaft's position, speed, or sometimes both. This signal feeds your distributorless ignition system and energizes your coils, causing them to fire.

Figures 6-63 and 6-64 are screens associated with monitoring a Ford's MAP sensor. The first illustration shows the MAP sensor with no vacuum applied. The second illustration shows how it looks when 20 psi is applied. Incidentally, regular ambient air pressure is about 14.7 psi. Note that the scales in both these screens are not the same (the time bases differ.) Also, note that these waveforms describing pressure applied are still symmetrical; however, their frequency decreases in proportion to this applied pressure. That is why it is important to look at the time base to get an accurate idea of just how much this square wave's frequency does decrease.

Figure 6-65 shows GM's optical sensor distributor pulses, with eight pulses (used as a reference signal). The pulse widths of these signals vary due to a varying position of the distributor. This is important information sent to your car's computer so that it can know which cylinder is firing.

Figure 6-66 shows a GM IAC control module signal. In this instance, a stepper motor actuates with its almost inherent small amount of generated electrical noise. This shows up as a small AC signal riding on top of a DC signal (ripple). This noise is from the brushes within a motor, just in case you wish to trace this back to the likely culprit. This is precisely the shape of a pulsed waveform your car's computer wants to see. When it is correct like this, it then can properly regulate airflow at idle. More specifically, this signal prompts or moves the pintle in and out of its seating in your car's throttle body assembly.

Figure 6-67 shows an example of a variable frequency pulse train that also has a variable pulse width. This signal comes from a GM feedback carburetor in which the pulse width controls the fuel and air bleed operation. As this pulse's width increases, it indicates a progres-

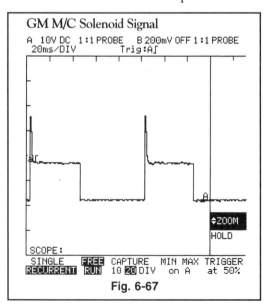

Fig. 6-67

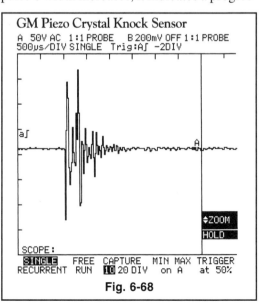

Fig. 6-68

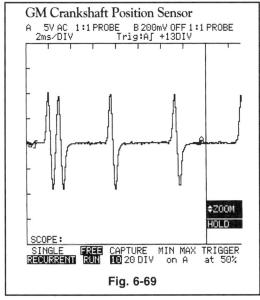

Fig. 6-69

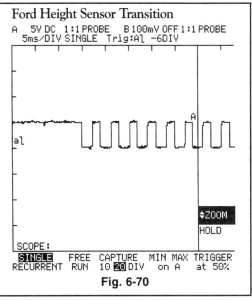

Fig. 6-70

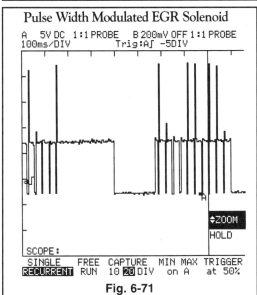

Fig. 6-71

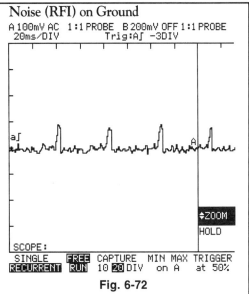

Fig. 6-72

sively leaner air-to-fuel ratio of flow. The spikes you see in this waveform are from the sudden discharge of the coil within the solenoid, since it is dumping a lot of energy off within a small amount of time.

Figure 6-68 is an unusual appearing waveform whose output signal is "damped." Damping is when a device oscillates with progressive loss of energy on successive cycles until it simply goes away and is no longer a concern. This is another case of a loss of and dumping of electrical energy. This particular example is a GM piezo crystal knock sensor. This crystal detects knocks, and once detected, sends your car's computer a signal to retard its timing.

Figure 6-69 shows the output of a GM crankshaft position sensor. The purpose of this sensor is to provide an input to your car's computer, indicating both the RPM and position of your car's rotating crankshaft.

Some modern cars have ultra-sophisticated suspension systems for not only a smoother ride, but also for purposes of safety. The sensor output in Figure 6-70 is that of a Ford, and it is a height sensor. How it works is that as your suspension system moves, it sends a signal to the computer. That, in turn, commands more or less pressure to be applied to the car's air suspension.

Figure 6-71 is admittedly a strange-looking signal, but it should not alarm you. It comes from your EGR solenoid, which is yet another example of a device that is PWM'd or pulse-width modulated or controlled. As its name implies, the pulse width of this signal controls the operation of this and many solenoids within a typical modern car.

As previously stated, EMI or RFI can be a real problem and source of nuisance, since it is especially prone to adversely affect your car's radio (and even cellular phones). Figure 6-72 shows this unwanted hash or electrical noise again riding on top of DC steady-state voltage. In this case, this RFI is about 500 mV or ½ volt in magnitude. There can be a number of sources of this hash, most of which emanate from a secondary ignition wire that can be physically residing too close to a semiconductor circuit. Also, other sources are where magnetic fields build up and collapse, characteristic relays, solenoids, and motors all possess. However, if you are an ambitious and really curious sort of reader, you can build what is called a "sniffer" probe. This device (see Figure 6-73), "sniffs out" RFI (also called EMI). Once built, you place your scope's probe tip within the open circular end of this pin. You will need to experiment a bit with changing values of components to get it low enough to where it detects the RFI in your car's major offending frequency. The larger the capacitor you use, the lower the frequency it can detect.

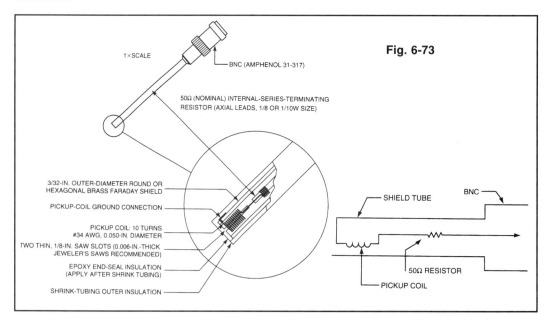

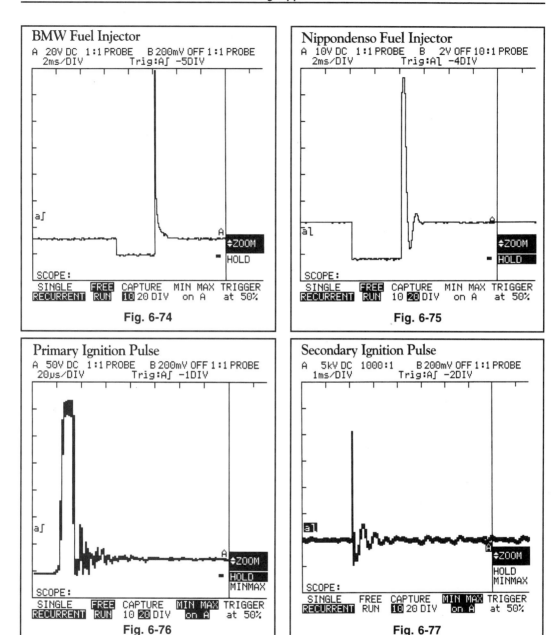

Figure 6-74 and 6-75 examine two very different types of fuel injectors. The first is from a BWM and what characterizes this waveform is that it saturates. That is, the transistor is driven on all the way. This engages the injectors and when the transistor is released from its saturated state, it shows a spike upward. This again, like in many previous examples, is an indication of free energy being suddenly liberated. Figure 6-75 specifically is an example of a complex waveform. This one comes from the familiar primary ignition coil's pulses. This is the familiar PWM scheme working here again. You may have seen this on car engine analyzers before. This signal shows considerable ringing (damped oscillations), which is normal since it comes from an ignition coil's pulse. In turn, this is induced from the collapsing and building up magnetic fields associated with this coil. Incidentally, the Fluke model 97 DSO we are now examining can perform this test more accurately than this older instrument, as well as many other tests it cannot perform. And it also does all this for a lot less money.

Figure 6-77 is the same general complex waveform and type; however, in this case it comes from a secondary ignition coil's pulse. Be cautious when measuring both primary and secondary coils since they generate voltages well into the thousands of volts. Understandably then, you will have to limit this large voltage since no instrument can accept it. The way to do this is with a 1,000:1 probe. This means that the probe absorbs 999/1,000 of the signal and only passes 1/1,000 of the signal to the scope's input. At 5 kV, this is still a rather healthy 5 Volts, even after the probe divides the signal down by 1,000. Figure 6-77 was taken with one of these 1,000:1 probes.

What's Next?

The following chapter concludes this book and examines your car computer and the error codes it generates. Be forewarned, if the CPU is faulty, it is impossible to fix. This is because a car's computer from the factory is loaded with codes that are not available to the general public. You may be able to fix some associated circuitry around the computer, but not the computer itself once it goes bad. But isolating down to this level is, in itself, a good troubleshooting task. This chapter also serves as a catchall in case we previously missed something.

Hall Effect Bibliography

Hall Effect Literature

1. Decesari, Rogert J., "Simple Hall-Effect Tachometer Measures Rotational Speeds Directly in RPS or RPM" Electronic Design p. 15, April 12, 1980.
2. Equizabal, Antonio L., "Hall Effect Tachometer Senses Speed, Direction of Rotation" Electronics p. 176 August 28, 1980.
3. Karasz, John, "Hall Sensors and Flip-Flop Sustain Pendulum's Swing" Electronics p. 131 May 22, 1980.
4. Lawrence, George L., "Hall Effect Magnetometer" Electronic Experimenter's Handbook 197 Edition, p. 48, 1977.
5. Malvar, Henrique Sarmento, "One-chip Tachometer Simplifies Motor Controller" Electronics p. 96 December 18, 1980.
6. Netzer, Yashay, "Bias Hall Sensors for Minimum Drift" EDN p. 180 March 17, 1982
7. Normet, Henno, "Hall-Probe Adapter Converts DMM into Gaussmeter" Electronics p. 171 January 3, 1980.

8. Polczynski, Mark, "Hall-Effect Sensor Provides Low-Cost, Reliable Isolation" Electronic Design p. 271 January 8, 1981.

9. Steinbaugh, Gary, "Hall Compass Points Digitally to Headings" Electronics p. 112 December 18, l980.

10. Micro Switch, a Division of Honeywell, "Specifier's Guide for Solid State Sensors" pp. 50 - 62, no date.

11. Sprague Engineering Bulletin 27701 "Hall Effect IC Application Guide", no date.

Addresses of Companies

American Electronic Components, Inc.
1010 North Main Street
P.O. Box 280
Elkhart, IN 46515
(219) 264-1116

Fluke Corp.
P.O. Box 9090
Everett, WA 98206
1-800-443-5853

Sprague Electric Company
70 Pembroke Road
Concord, NH 03301
603-224-1961

Tektronix, Inc.
P.O. Box 500
Beaverton, OR
97077-0001
503-627-7111

Chapter 6 Quiz

True or False

1. A sine wave is produced by a DC source.
2. Sine waves are produced from an electrical device that generates a signal with constantly varying amplitude with respect to a stationary object; therefore, it tends to recycle and rotate.
3. An example of an automotive device producing sine waves is a permanent magnet type of vehicle speed sensor.
4. Frequency is expressed in terms of voltage and current combined.
5. The computer based devices within your car can receive and intelligently decipher sine waves.
6. Square waves are symmetrical; therefore, they have equal or 50 percent duty cycles.
7. A pulse train is not symmetrical.
8. Pulse trains are most often derived or made from sine waves.
9. Drivers on electronic fuel injectors provide a path to ground for current flow.
10. The distance a solenoid moves against its return spring is not at all related to the average current flowing through its windings.

Matching

11. These electronic components tend to distort waveforms and cause voltage spikes.
12. These devices often assume the physical form of two layers of a material that is a conductor and sandwiched between these layers is an insulator (dielectric).
13. An IC that contains four power or rather robust transistors within its case is called one of these.
14. The regulator controls the motor current in a CS series generator by this.
15. This electronic component often is used to conduct voltage and current harmlessly to ground.
16. Electric motors sometimes have their brushes experience bouncing and arcing due to the field controlling it only partially doing this.
17. The average current within any device controlled by the PWM method is proportional to the _____ percentage of time the voltage is in it active state.
18. In order for inductance to exist there must be a current passing through an conductor and this conductor must physically be _____.
19. A _____ responds in direct proportion to the average current flowing through its windings.
20. These devices are less prone to clog and are meeting with great widespread success in more modern cars.
21. There are two types of electronic fuel injectors, the throttle body injector (TBI(c) and this type.
22. These devices more precisely determine and participate in the dispensing and ultimately in determining the optimum air-to-fuel ratio. The result it that they deliver the most power per unit of fuel used.
23. The main difference between the two types of electronic fuel injectors is their numbers and the position in which they are _____.
24. This is a condition, most often when excessive heat is present, in which the transistor continues to pass current, even when conducted to shut off.
25. Capacitors and inductors are often combined to form one of these, which defeats electrical noise that otherwise might interfere with car radios and cellular phones.

A. PWM
B. PERCENTAGE
C. DC ELECTRIC MOTOR
D. INDUCTORS
E. COILED
F. PARTIALLY COLLAPSING
G. CAPACITORS

H. DIODE
I. QUAD DRIVER
J. CHOKE
L. THERMAL RUNAWAY
M. FUEL INJECTORS
N. PORT FUEL INJECTOR
O. PLACED
P. BALL VALVES

7

Your Car's Computer

This last chapter examines your car's computer. As you will discover, it plays a vital role in helping you more exactly and rapidly troubleshoot faults. This chapter also shows how your car's computer is miraculously designed to keep many major systems just purring and optimized for maximum performance and minimum fuel consumption. This "fine tuning" that your car's computer does is only possible by it having the ability to read sensor outputs. It uses this sensor data in making a comparison against a known good value. These values are located in a repository within your car's computer (where many other fault codes also reside).

Thus far we have seen how you analyze numerous faults. The fault indications have all originated from sensors within your car. We have used various types of electronic test instruments to examine waveforms and monitor voltage levels. However, ask yourself, "Why do these sensors all have outputs?" Sure, they allow monitoring with your meter or scope. But the real answer is that these sensor outputs primarily are there to feed your car's on-board computer.

Your Car Computer's Responsibilities

As your car's computer receives this data, it properly formats it, determines which sensor the data is coming from, and makes a comparison with what it expects to see. If these do not match, then you have a fault. This assumes that the sensor's output is 100 percent functional. That's another good reason why your car's computer monitors these sensors (or actually they pass data back-and-forth). Figure 7-1 shows a typical car's computer, sensors, actuators, and test connectors. Note that there are three test connectors. One is located at top center and the other two are at the lower right and upper left of Fig. 7-1. By the way, sensors that obviously transform one type of energy into another are often called transducers. An example might be a temperature-to-voltage converting integrating circuit.

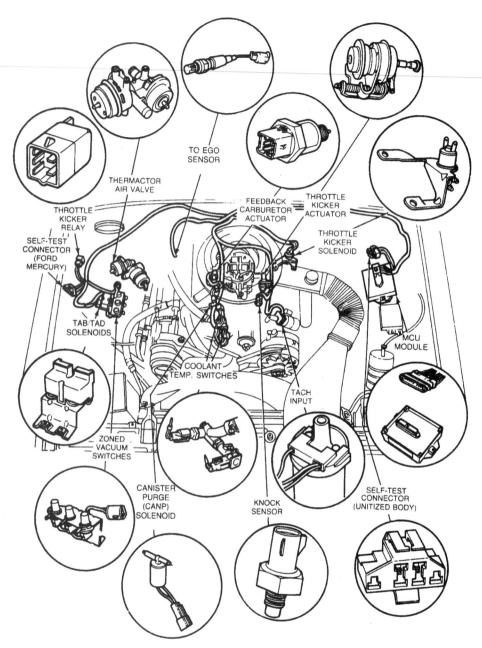

THERMACTOR
AIR VALVE

TO EGO
SENSOR

FEEDBACK
CARBURETOR
ACTUATOR

THROTTLE
KICKER
ACTUATOR

THROTTLE
KICKER
RELAY

THROTTLE
KICKER
SOLENOID

SELF-TEST
CONNECTOR
(FORD
MERCURY)

TAB/TAD
SOLENOIDS

MCU
MODULE

COOLANT
TEMP. SWITCHES

TACH
INPUT

ZONED
VACUUM
SWITCHES

CANISTER
PURGE
(CANP)
SOLENOID

KNOCK
SENSOR

SELF-TEST
CONNECTOR
(UNITIZED BODY)

This overall view of a Ford computerized engine control system shows the relationship of the computer (MCU, or module), information sensors and output actuators

Fig. 7-1
Typical car's computer, sensors, actuators and test connectors

Control Obtained Through Feedback

There are many systems within your car that your computer also attempts to control. A lean-burning carburetor is one such system. As your carburetor makes a richer or leaner mixture of fuel-to-air, the computer senses if the carburetor's decision is correct. That is, did it improve the situation by bringing it back into a range of correct mixture, or did it make it worse? Your car's computer makes this decision by using *feedback controls*. And it is precisely this feedback control that makes your car's computer so useful.

You can most easily comprehend this concept of feedback control by thinking of the thermostat within your home's heating and cooling system. In older models there was a bi-metallic strip of metal. Two dissimilar strips of metals are welded together to form this strip. Since these metals have different coefficients of expansion, one expands at a rate greater than the other. This makes the strip bend or straighten out. As your home heats, it bends and touches a metal switch to shut off the heating unit. As your home cools the strip straightens out and pulls away from this metal switch. This restarts your heating and/or cooling unit. Figure 7-2 is another example of a thermostat, only one that resides in your car's cooling system. This very simple example of a feedback control system basically closes and opens the thermostat in response to the temperature of the coolant within your radiator. The opening and closing are precisely analogous to the bending of the bi-metallic strip.

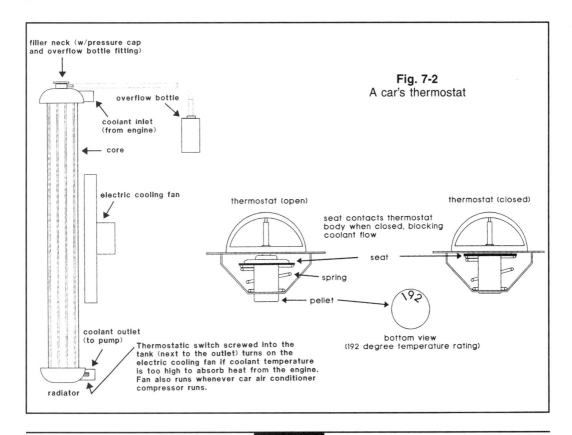

Fig. 7-2
A car's thermostat

filler neck (w/pressure cap and overflow bottle fitting)

overflow bottle

coolant inlet (from engine)

core

electric cooling fan

thermostat (open)

seat contacts thermostat body when closed, blocking coolant flow

seat

spring

pellet

thermostat (closed)

bottom view
(192 degree temperature rating)

coolant outlet (to pump)

Thermostatic switch screwed into the tank (next to the outlet) turns on the electric cooling fan if coolant temperature is too high to absorb heat from the engine. Fan also runs whenever car air conditioner compressor runs.

radiator

Error Codes

To this point in our troubleshooting examples, we have gone right to the car's problem within the electrical system. But this is rarely the case in the real world. Almost no experienced master mechanic can even be that smart (or lucky). You will first have to fault isolate (troubleshoot) a general or, better yet, a specific area of trouble. This is where the computer in your car shines once again. It generates fault or error codes.

Standard, and Not-So-Standard Codes

Each of these codes corresponds to a different problem. However, these codes are sometimes manufacturer specific and often even model and model-year unique within cars even made by the same company, such as GM and Ford. This is especially true of the very earliest cars using on-board computers. But today there is a very concentrated effort to standardize these codes.

Equipment Useful on more than One Type of Car

That allows a test equipment manufacturer to design a diagnostic instrument for car computers that will be useable with far more types of cars. There still exists, though, cars that tend to have features unique only to that model or brand. Obviously, in these cases, the very nature of them only existing on one car makes them unique—with a corresponding unique code. An example might be a car whose hardtop roof retracts and an error code indicates it is jammed. Be forewarned though, a new Dodge Avenger may not share all the same error codes that a two-year-old Buick Riviera might have. Again, this is simply because each car has some unique features.

Practical Troubleshooting Associated with Car Computers

You, nor even the dealership, can fix your car's computer if it proves to be defective. It is more expedient in time and money just to replace the whole assembly. There is a printed circuit (PC) board upon which the computer and its components reside. A PC board is a thin wiring pattern composed of fine lines made of very thin copper foil on these boards. They serve mainly as interconnecting electrical paths between the components.

"Printed" Wiring = Higher Reliability

This saves stringing numerous wires all over the space under the dash and hood in order to connect them all. Look at wiring harnesses of older cars and see the bundles of wires neatly bundled and tied at about 8-inch intervals with string. This labor-intensive task greatly increased the cost of older cars and lowered their reliability as well. Contrast this to a modern car that uses a thin, very pliable and flexible PC board that often is long and narrow so it can fit all the way across the space occupied under your dash. No wires exist here, just copper foil runs.

Fig. 7-3

Also, some modern car computers have special sockets that hold the chip or integrated circuits (see Fig. 7-3). Unfortunately, most of the microprocessor ICs are just soldered onto the PCB (Printed Circuit Board). That leaves you with no alternative other than to treat this assembly as an unrepairable throwaway part. But in this author's 1991 Pontiac Grand Prix this assembly costs $108 in the after-market, which is not that much for such a workhorse of a part. The term after-market means a part that is made to match the original manufacturers part, but at a much smaller price.

The Achilles Heel

Generally, you will discover that the energy management, fuel injection, and emissions controls are the systems most actively managed by your car's computer. You can be assured that you will encounter far more problems with wiring (which a PC board minimizes), vacuum hoses, and connections made in any manner (switch, terminal block, splice, etc.). Other sinister factors contributing to reducing the useful life of your car and lowering the average time between electrical and mechanical failures are rust, oxidation, and corrosion. These specifically offer paths of greater resistance to the very small currents driving and originating from sensors.

Surprise, Surprise!

You car's computer does not expect to see this added resistance. And when it does encounter it, it compares a signal to a memory look-up table and may well send a fail code out. Other factors compounding the difficulty of this situation are heat, moisture, vibration, and corrosive salt, and these generally become progressively worse as your car ages and becomes "looser" with many mechanical and electrical joints that have been stressed thousands of times.

Getting Started

Let's begin with your fuel management system. But first things first; you will have to determine if there is some small quirk that makes it not function properly. Otherwise, the electronic controls have inaccurate information to manage the engine fuel injection systems.

Immediately reverting to cursing your car's computer (and its input sensors and output actuators) may lower your blood pressure but it leads to an inaccurate diagnosis.

Spotting "Quirks" that Can "Taint" Data

Table 7-1 is a helpful overview of how you should proceed so your measurements are meaningful and can be trusted. Otherwise, some interacting system that has no business of interacting with your engine management system might "taint" your car's on-board computer diagnosis and eventually the error code it produces. The portion about checking the discharging system is crucial.

Table 7-1

Actions to Take First

- Determine that your engine is basically sound by making compression and vacuum tests—followed by checking all hoses, even ensuring that each hose goes to its proper place.

- Inspect if your battery and all electrical connections are free of corrosion, that the discharging and discharging systems is okay, and that all fuses and fusible links are intact.

- Check to determine if the fuel and air supplies are free of obstructions or object that may potentially block their inputs.

- Perform a very quick overview check on the EGR, PCV and EVAP emission controls—followed by determining that the ignition system is also functioning properly. This means ensuring that you'll avoid any cross-firing or "carbon tracts."

- Lastly, make certain that your car's computer is in the closed loop mode of operation.

What to Expect After Changing Batteries

There are precautions to follow, the most notable of which is when you disconnect your battery. This makes your computer vulnerable to losing its stored data. It stores this data that you can lose in RAM (Random Access Memory). After you first install a new battery you may often experience driving problems. However, this is only temporary.

Once your car's computer completes its "relearn" cycle, everything should go very smoothly. Also, more modern cars now have stereos and radios with antitheft features. Make certain you have the correct activation code before disconnecting your battery. Table 7-2 shows a listing of fault codes for such a situation with a late-model six-cylinder Chrysler product. Note here the codes "11" and "12," which specifically pertain to this case.

Table 7-2

Computer trouble codes

Code	Probable cause
64	Flexible fuel (methanol) sensor indicates concentration sensor input less than the acceptable voltage
65	Manifold tune valve solenoid circuit open or shorted - On 1996 models, power steering switch failure
66	No message from the transmission control module (TCM) to the powertrain control module (PCM)
66	No message from the body control module (BCM) to the powertrain control module (PCM)
71	5-volt PCM output low
72	Catalytic converter efficiency failure
77	Speed control power relay circuit
88	Start of test
11 (Dakota pick-up, 2.5L models)	Engine not cranked since battery was disconnected
11	Engine not cranked since battery was disconnected/no distributor input signal
12	Memory standby power lost
13*	MAP (Manifold Absolute Pressure) sensor vacuum circuit - slow or no change in MAP sensor input and/or output
14*	MAP (Manifold Absolute Pressure) sensor electrical circuit - high or low voltage
15**	Vehicle speed/distance sensor circuit
16*	Loss of battery voltage

* These codes ligt up CHECK ENGINE light
** These codes light up CHECK ENGINE light on vehicles with special California emissions controls

Safely Handling Batteries

A battery is a reservoir of tremendous energy. In order to turn over your car's cold engine, it has to be able to release this vast energy over a rather short time (measured in seconds). One way to have something that will absorb this energy is to use a carbon thermopile, see Fig. 7-4. As its name implies, it is a stack or pile of carbon discs that are specifically designed and connected to absorb great amounts of energy over a very small time interval.

Trying to use a dead, uncharged battery as a source for downloading or discharging a good battery can prove lethal! A battery can only be charged at a certain specified rate of energy transfer. This rate is much slower than its nearly instantaneous discharge rate when you try to charge an uncharged battery with a fully charged one. This is hazardous since it severely electrically stresses the uncharged battery's plates, which can almost instantly buckle under

Fig. 7-4
A thermopile energy absorber
or load

the electrical strain—and it subsequently explodes. Since you might be in a hurry, using a
thermopile to accept this sudden slug of energy is a safe method of discharge.

Closed Loop vs. Open Loop

There are two modes in which your car's computer runs. These are the *closed-loop* and
the *open-loop* modes, and it is vitally important that the on-board computer be able to get from
the open-loop to the closed-loop modes so it can satisfactorily perform its monitoring job.

The open-loop mode occurs when you first start your car and the engine and its all-
important oxygen sensor have yet to heat up and stabilize themselves. Until all of the factors
involved in your car's computer stabilize, it is difficult to make an intelligent troubleshooting
decision. The computer remains in the open-loop mode until everything stabilizes. In practical
terms, this means that all positions or settings stay "fixed" at the levels that existed when the
car first left the factory.

The closed-loop mode of operation is the normal mode in which your computer almost always operates. Even if your car's engine and oxygen sensors are already stabi-
lized, your car's computer will still wait for a little while before entering the closed-loop
mode of operation.

Retrieving Trouble Codes from Your Car's On-Board Computer

There are many ways this is done, and mostly it depends on what company manufac-
tured your car. Typically, most cars have lights on the dashboard (the so-called idiot lights).
These are accompanied by a legend such as "CHECK ENGINE", "POWER LOSS", "SER-
VICE ENGINE SOON", etc. You can use these lights to blink the codes that are stored
within your car's computer when you manually trigger your computer through its diagnostic
connector.

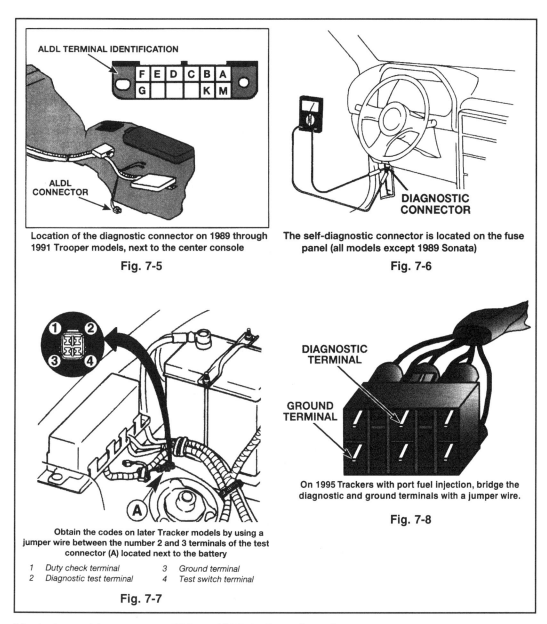

Location of the diagnostic connector on 1989 through 1991 Trooper models, next to the center console

Fig. 7-5

The self-diagnostic connector is located on the fuse panel (all models except 1989 Sonata)

Fig. 7-6

On 1995 Trackers with port fuel injection, bridge the diagnostic and ground terminals with a jumper wire.

Fig. 7-8

Obtain the codes on later Tracker models by using a jumper wire between the number 2 and 3 terminals of the test connector (A) located next to the battery

1	Duty check terminal	3	Ground terminal
2	Diagnostic test terminal	4	Test switch terminal

Fig. 7-7

Variations, Variations—When Will It Ever Stop?

Figures 7-5 and 7-6 show two variations of where the test connector is placed within a car. To further demonstrate this total lack of standardization in testing by your car's computer, see Figs. 7-7 and 7-8. Figure 7-7 shows how a late-model Tracker with ported fuel injection differs from an identical Tracker without ported fuel injection. If this overwhelming data hasn't already convinced you, Fig. 7-9 shows how Hyundai's Scoupe and Accent even have differing styles of test plugs (connectors).

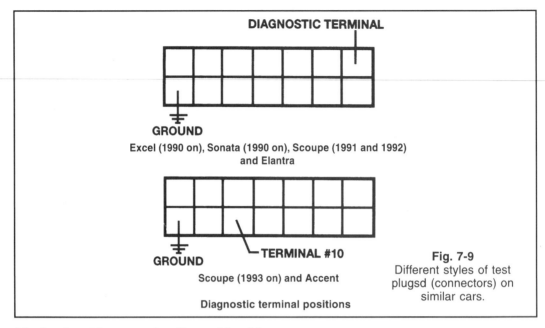

DIAGNOSTIC TERMINAL

GROUND

Excel (1990 on), Sonata (1990 on), Scoupe (1991 and 1992) and Elantra

TERMINAL #10

GROUND

Scoupe (1993 on) and Accent

Diagnostic terminal positions

Fig. 7-9
Different styles of test plugsd (connectors) on similar cars.

Methods of Interpreting Data Also Vary

On other cars you connect a voltmeter to the diagnostic connector and start counting the needle swings (this is only on an analog meter) or you can also observe an LED readout on the computer's board (assembly) itself. Regardless of how your car allows you to obtain or gain access to these codes, you will have to check them against the chart that pertains to your specific model of car. Figure 7-10 shows a typical LED (light emitting diode) display. It consists of seven segments, each of which is a separate LED (see Fig. 7-11). You therefore can activate various combinations of segments to form numerals. For example, activating all segments results in an "8."

Fig. 7-10
Discrete stand-alone LED display

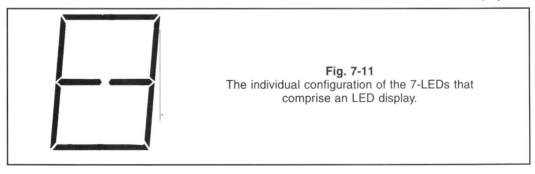

Fig. 7-11
The individual configuration of the 7-LEDs that comprise an LED display.

The OBD II System

As previously stated, recognizing the folly of early cars with on-board computers, there is an ongoing effort to standardize diagnostic fault codes. The OBD (**On**Board **D**iagnostic) system was the first to attempt standardization. The convention adopted was a letter followed by a four-digit numeral. The OBD requirement is now so massive that it is federal law to have such a system on all new cars built in America or imported to America. Also, many states now even have a program whereby you can drive in a "wreck" and get paid a set sum for it. This is because 90 percent of all the pollution is caused by just 10 percent of all cars on the road. Those 90 percent are also the oldest cars running.

On-Board Computer or Not?

How would you know if you had a car that was just on the edge of cars with and without on-board computers? It's simple; just look at the VECI decal on top of the radiator fan shroud. It should say "OBD certified." Figure 7-12 shows one of these VECI decals. Let's never lose sight of the primary goal and driving force behind this whole computer car and regulated and controlled industry, as we know it so very well today! It resulted from stricter standards concerning emission controls and improved economy. The spin-offs that accompanied these first efforts now dramatically influence the present and future designs in cars.

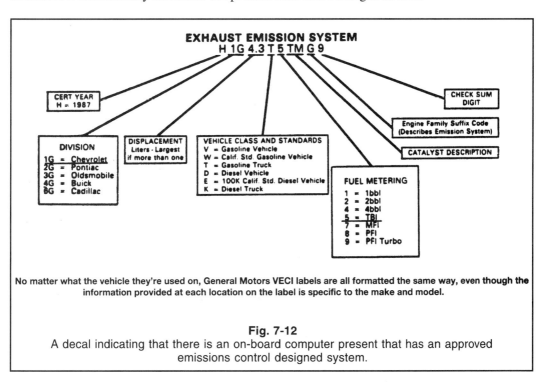

No matter what the vehicle they're used on, General Motors VECI labels are all formatted the same way, even though the information provided at each location on the label is specific to the make and model.

Fig. 7-12
A decal indicating that there is an on-board computer present that has an approved emissions control designed system.

Troubleshooting Your Fuel Pump

As our next example, let's troubleshoot your car's fuel pump. If your engine cranks (the battery turns it over), but it won't start, the likely culprit is the fuel pump. Get a buddy and have him turn your ignition key ON and you should observe the fuel pump trying to activate. If you don't hear anything, then it is not getting current since it is an electric pump. Also, fundamental to this discussion, you must know that it requires three elements for a car to run. These are fuel, spark, and compression. If any one of these is missing then the car will not start.

Figure 7-13 is an illustration of a typical fuel pump relay, fuse, and fuel pump circuit. Table 7-2 shows the fuel pump fault that your car's computer displays, and it is "64" for a six-cylinder recent Chrysler product. Figure 7-14 shows what the schematic in Fig. 7-13 looks like, first with all wiring in place. Next, referring back to Fig. 7-1, you will see an "exploded-view drawing." This is complete with all of the "breakouts" for allowing you to gain access to place them where you can test the computer's output signals.

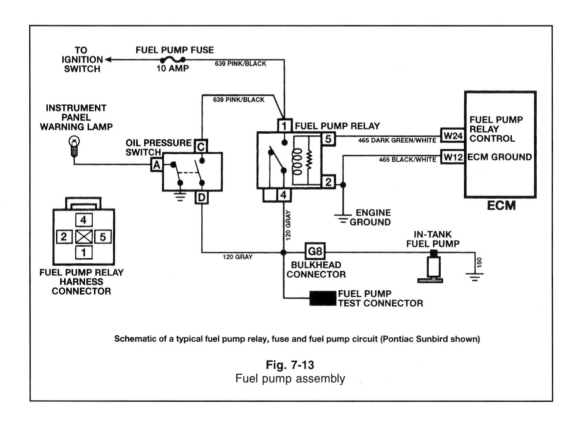

Schematic of a typical fuel pump relay, fuse and fuel pump circuit (Pontiac Sunbird shown)

Fig. 7-13
Fuel pump assembly

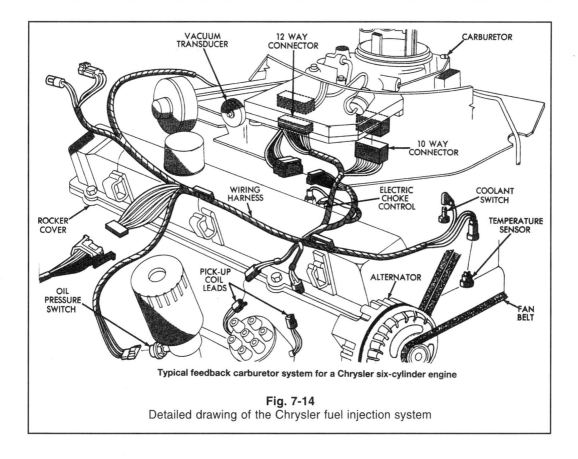

Typical feedback carburetor system for a Chrysler six-cylinder engine

Fig. 7-14
Detailed drawing of the Chrysler fuel injection system

Referring back to these error codes in Table 7-2, note several more interesting occurrences. One is that there are four error codes with asterisks beside them. These correspond to activating a light on your dash (the idiot light) when they become active. Also, the error code with a double asterisk, namely 15, corresponds to cars made for use in California. As an example, Mercedes Benz modifies its diesels in order to pass California's Clean Air Law so they can be marketed there. The other item is code 12. This is active after you have changed your car's battery and your car's computer has to "relearn" what it is supposed to do with these codes.

If the fuel pump fails to make any noise or sound in its attempt to actuate, the current that should be flowing through it must not be flowing. This means that there is an opening in the circuit. This usually is a broken wire or a blown fuse. A fuse is a device best thought of as the "weakest link by design" in a circuit. If you design a circuit to carry ¾ Amperes you might place a 1-Ampere fuse in line (in series) with it. If the current exceeds 1 Ampere then the fuse blows.

But imagine what would happen if the fuse either refused to blow or were not there at all. You would have a great surge and magnitude of current flowing and that surely would damage delicate components not designed to handle such robust amounts of current flow.

The Fuse: Very Important, Yet so Simple

Figure 7-15 shows what a typical fuse looks like in a modern car. If you have ever held a tungsten light bulb up to the light and noticed the very fragile thin filament, then you can appreciate a fuse. Specifically, on a burnt light bulb you will see that the wire has opened up or ruptured — just like what occurs in a fuse. A quick and safe way to test for fuses such as the one in Fig. 7-15 is to use your DMM and perform a continuity test. This is where the beeper feature on a DMM comes in handy in a dark garage or at night.

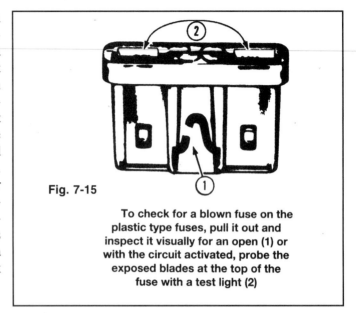

Fig. 7-15

To check for a blown fuse on the plastic type fuses, pull it out and inspect it visually for an open (1) or with the circuit activated, probe the exposed blades at the top of the fuse with a test light (2)

Older "Conventional" Fuses

There are also more conventional-appearing fuses (see Fig. 7-16). These type fuses have a thin "ribbon" filament stretching from one end to the other. When the circuit in which they exist experiences an excessive amount of current, the fuse's filament ruptures or opens.

Fig. 7-16
Traditional miniature fuse with its holder

This prevents flow of what would otherwise be an excessive and ultimately destructive amount of current.

Many a misguided "wannabe" mechanic has faced the blowing fuse problem by using "brute force" rather than the more tedious method of using analysis. Specifically, putting in a far larger current rating fuse results in burning out complete electrical wiring systems in cars, so be forewarned about this less analytical "temperamental" approach to troubleshooting!

Back to Your Car's Feedback Carburetor System

This is a far more complicated system than a mere fuel pump (which, granted, is an integral part of this system). Refer back to Fig. 7-14, which shows such a system, and note the sensors and electrical components on this Chrysler Division's six-cylinder car. Table 7-2 shows some typical fault codes on a late-model six-cylinder Chrysler car. The trouble code for this situation is "27."

A New Concept and Type of Data

Since these codes come from your car's computer, you can't view them as we previously did with your car's sensors. These codes are expressed in binary codes. The term binary means that there are two states or only two types of numbers by which you express all numbers within this numbering system. These are a binary 1 and a binary 0. Before we start to examine the binary numbering system, let's step back and take (maybe for the first time ever) a really hard close-up look at our own decimal numbering system.

Some tasks that we do in life we do so often that they become second nature to us. That being the case, most of us never really stop to even think about them very much—much less ever go to the trouble of analyzing them in-depth. Such is the case with our own numbering system. To give you an idea of how this works, let's look at how we write numbers in our ordinary everyday system called the decimal system.

The term decimal means *ten*, and that is precisely how many choices we have in the form of different numbers. These decimal digits are naturally: 9, 8, 7, 6, 5, 4, 3, 2, 1, and 0. If it isn't already apparent to you, the decimal system allows you to express larger numbers by using fewer digits. If we want to express the number 3,814 in decimal (see Fig. 7-17), what we are actually doing is taking the number to the far left (the 3) and multiplying it by (1,000). You multiply the next number after that to the right (8) by 100. Again, you multiply the next number after that to the right (1) by 10. Finally, you multiply the last digit to the far right (4) by 1.

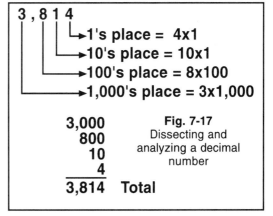

Fig. 7-17
Dissecting and analyzing a decimal number

```
3,814
     └→1's place = 4x1
     └→10's place = 10x1
     →100's place = 8x100
     →1,000's place = 3x1,000

3,000
  800
   10
    4
3,814  Total
```

When you add the sum of these four multiplication operations, you arrive at the answer. By accepted convention, the digit to the far left is called the *MSB*, (most significant bit) since it has the greatest impact in determining the final overall number. The digit to the far right is called the *LSB*, (least significant bit) since it has the least impact in determining the final overall number.

Now that we know how decimal numbers are actually composed, the most important lessons learned, which are directly applicable to our attempts to learn the binary system, are

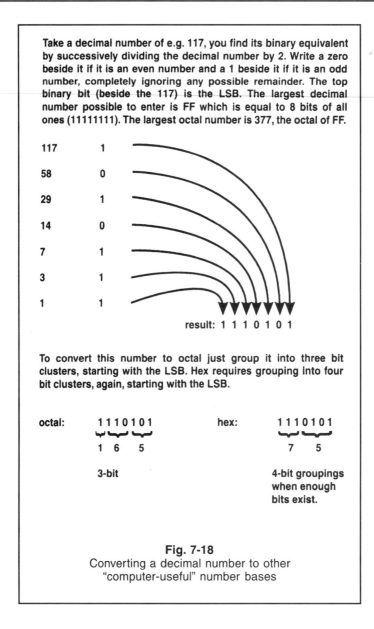

Take a decimal number of e.g. 117, you find its binary equivalent by successively dividing the decimal number by 2. Write a zero beside it if it is an even number and a 1 beside it if it is an odd number, completely ignoring any possible remainder. The top binary bit (beside the 117) is the LSB. The largest decimal number possible to enter is FF which is equal to 8 bits of all ones (11111111). The largest octal number is 377, the octal of FF.

117	1
58	0
29	1
14	0
7	1
3	1
1	1

result: 1 1 1 0 1 0 1

To convert this number to octal just group it into three bit clusters, starting with the LSB. Hex requires grouping into four bit clusters, again, starting with the LSB.

octal: 1 1 1 0 1 0 1   hex: 1 1 1 0 1 0 1
        1   6   5   7   5

3-bit   4-bit groupings
                                 when enough
                                 bits exist.

Fig. 7-18
Converting a decimal number to other
"computer-useful" number bases

two-fold. The first lesson is that the numbers to the left of each adjacent number are greater in magnitude. Also, you multiply each successive digit by 10, then by the actual decimal numeral itself (0 to 9). As an example, refer again to Fig. 7-17. The number we used, 3,814, has its 8 in the 100's place and the 1 in the 10's place. That means that it is 10 times greater than any digit in the 10's place. After this 10 assessment you have to multiply the number by what numeral appears there, such as the 8 and the 1.

In the binary numbering system, there are only two number choices—0 and 1. Therefore, as you go across from right to left, each successive place is twice as great as the last (instead of 10 times as great)! So the right most digit, the LSB is equal to 1, then the next is 1x2 = 2, then the next is 2x2 = 4, then next is 4x2 = 8, the next is 8x2 = 16, 32, 64, 128, 256, 512 ….. But to use this system, let's first examine the binary number 101. Starting from the right it is 1x1 = 1, added to 2 x 0 = 0, added to 1 x 4 = 4, so it is 4 +1 = 5 in decimal. Figure 7-18 shows how to take a decimal number and convert it to binary, octal, and hex numbers. Octal is to the base 8 and hex (hexadecimal) is to the base 16. That would mean that you could express larger numbers in hexadecimal than you could in the decimal numbering system.

Wrapping It All Up

Certainly, you now have an excellent idea of what your car's computer can and does do. The toughest part, though, may well be what appears like the simplest part. That is, to find where the computer's output fault code connector is situated. In case you ever want to use another more sophisticated instrument, a logic analyzer, you will have to understand the binary codes since it displays all binary ones and zeros. You can also display data in octal and hexadecimal as well. But that should not be a problem after your crash "minicourse" on binary numbering systems presented earlier.

Also, if you were intrigued by what was presented here, you may wish to more deeply delve into this subject. If you do, you will inevitably use far more sophisticated test instruments; therefore, knowing the binary system and number base conversions will greatly help you. Not knowing this will render you incapable of interpreting data presented in this number base. Good luck with all your future electrical system and computer troubleshooting on your cars!

Chapter 7 Quiz

True or False

1. Your car's computer attempts to maximize performance while minimizing fuel consumption.
2. Your car's computer passes data back and forth between its sensors and actuators.
3. The carburetor is one system on your car not controlled by your car's computer.
4. A thermostat works off the principle that two dissimilar pieces of metal expand and contract at the same rate.
5. Each code within your car's computer will be unquestionably the same on all models of cars.
6. There is a strong effort at present to standardize error codes on your car's computer.
7. One benefit of standardization is that an instrument can be designed to work on more than just one type of car.
8. Using a flexible PC assembly under the dash decreases reliability.

9. Small wires with insulation over them run all over the surface of a typical PC board.
10. Wiring in older cars tended to be more labor intensive and less reliable.

Fill in the Blanks

11. Generally ICs are soldered onto PC boards and not placed in _____.
12. The term for a part that is not made by the original manufacturer but is nearly identical to it is called an _____ _____ part.
13. The areas of trouble that are the most common on a car are the vacuum hoses, and _____.
14. Dirt, rust, corrosion and age all contribute to making a sensor's output signal have a higher _____.
15. Added resistance in signals can cause your car's computer to mistakenly interpret these signals as _____ _____.
16. A _____ is a tremendous reservoir of energy.
17. A _____ _____ is a mechanical device designed to absorb massive amounts of energy over short periods of time (or bursts.)
18. The _____ _____ of operation within your car's computer is when your engine and oxygen sensor are all warmed up and is ready to go.
19. The _____ _____ of operation within your car's computer is when your engine and oxygen sensor are both not all warmed up and therefore not yet ready to go.
20. One method of retrieving computer error codes is through lights on your car's dash, otherwise known as _____ lights.

Matching

21. Needle swing that follow a signal
22. Fuel, spark, and compression
23. Cause 90 percent of all car pollution
24. No noise from fuel pump
25. The whole push behind modern car designs
26. Segments form decimal numerals on an LED display
27. Tungsten ribbon opens or ruptures
28. A way of audibly detecting continuity in a series circuit
29. The OBD II system

30. VECI decal identification of this
A. Emission controls
B. Pollution and fuel economy
C. To start and run your car
D. Analog meter
E. On-board computer
F. Older cars
G. Beep
H. Seven
I. Fuse blows
J. Open circuit

Appendix

Quiz Answers

Chapter 1

1. F
2. T
3. F
4. F
5. T
6. F
7. F
8. T
9. F
10. T
11. E or H
12. C
13. I
14. J
15. E or H
16. G
17. B
18. F
19. C
20. D
21. D
22. D
23. D
24. D
25. C
26. D
27. C
28. B

Chapter 2

1. F
2. F
3. T
4. F
5. T
6. T
7. T
8. T
9. F
10. T
11. D
12. A
13. G
14. T
15. P
16. H
17. Q
18. L
19. J
20. R
21. S
22. O
23. I
24. C
25. N
26. K
27. B
28. E
29. M
30. F
31. short
32. rotating
33. test
34. 0.3 volts
35. voltage
36. noise
37. D
38. D
39. B
40. C
41. A
42. D
43. B
44. A
45. B
46. B
47. D
48. A

Chapter 3

1. T
2. T
3. F
4. F
5. F
6. T
7. T
8. T
9. F
10. F
11. 7
12. self-test
13. analog pointer
14. germanium
15. recharged
16. resistance
17. higher
18. inductive pickup
19. points
20. pulse width

Chapter 4

1. T
2. T
3. F
4. F
5. T
6. T
7. F
8. F
9. F
10. T
11. E
12. C
13. F or H
14. I
15. F or H
16. B
17. D
18. A
19. J
20. G
21. Internal and external
22. Vertically
23. Derived measurements
24. Heating
25. Lissajous
26. X-Y
27. Z
28. Amplifiers
29. Overshoot
30. 0.707

Chapter 5

1. T
2. F
3. T
4. T
5. T
6. F
7. F
8. F
9. T
10. F
11. G
12. V
13. A
14. O
15. J
16. D
17. Z
18. U
19. Q
20. M
21. E
22. T
23. Y
24. I
25. W
26. L
27. H
28. S
29. P
30. R
31. F
32. K
33. N
34. C
35. B

Chapter 6

1. F
2. T
3. T
4. F
5. F
6. T
7. T
8. F
9. T
10. F
11. D
12. G
13. I
14. A
15. H
16. F
17. B
18. E
19. C
20. P
21. N
22. M
23. O
24. L
25. J

Chapter 7

1. T
2. T
3. F
4. F
5. F
6. T
7. T
8. F
9. F
10. T
11. Sockets
12. After Market
13. Connectors or Connections
14. Resistance
15. Error or Fault
16. Battery
17. Carbon Pile
18. Closed Loop
19. Open Loop
20. Idiot
21. D
22. C
23. F
24. I or J
25. B
26. H
27. J
28. G
29. E
30. A

Reference

Acronyms Commonly Used in Automotive Electronics

4EA   Electronic automatic 4-speed transaxle

4X4L   4X4 low input switch

A4LD   Ford automotive 4-speed lock-up converter drive

AAC   Auxiliary Air Control Valve

AAV   Anti-Afterburner valve

ABS   Anti-Lock Braking System

A/C   Air conditioning; also can be alternating current

ACC   Air conditioning clutch compressor signal input to the computer relating status of the air conditioning clutch

ACCS   Air conditioning control switch

ACD   Air conditioner demand switch

AC DV   Air cleaner duct and valve control

A/C P   Air conditioner cutoff fan

ACP   Air conditioning pressure switch

ACT   Air charge temperature sensor or its signal output

ACV   Air control valve or thermostat air control valve

A/F   Air-to-Fuel ratio

AFC   Airflow controlled fuel injection

AFS   Airflow sensor

AHFSS   Air conditioning and heater function selection switch input to the computer

AI   Air injection

AIC   Automatic idling control valve

AIR   Air injection reaction system. Injects air into your exhaust system so your car burns all remaining fuel

AIR BPV   Thermactor air bypass valve

AIS   Air Injection System or automatic idle speed control and/or motor

AISC   Air induction injection system control. Like the AIR, it injects air into your exhaust system so your car burns all remaining fuel

AIV   Air injection valve

ANTI-BFV   Anti-backfire valve

AOC   Automatic overdrive transmission

ALU   Arithmetic Logic Unit

APC   Automatic performance controls

APS   Atmospheric pressure sensor

ASCD   Automatic speed control device (cruise control)

ASD   Automatic shutdown relay driver circuit (fuel pump relay)

A/T   Automatic transmission

ATDC   After top dead center

ATM   Automatic test mode

ATS   Air temperature sensor

AVOM   Analog volt/Ohm meter

AWG   American wire gage system,

smaller number wires have larger diameters. For each three AWG numbers you have a doubling or halving of the cross sectional area. This depends if you ascend or descend in AWG numbers.

BCD	Binary Coded Decimal
BDC	Bottom-Dead-Center
BSFC	Brake Specific Fuel Consumption
BTDC	Before Top-Dead-Center
CAFE	Corporate Average Fuel Economy
CAMS	Computerized Automotive Maintenance System
CC	Condition Code
CIS-KE	A type of airflow sensor made by Bosch
CKP	Crankshaft Position
CMP	Camshaft Position
CO	Carbon Monoxide, a poisonous gas
CPI	Central Port Injection
CPU	Central Processing Unit
CRT	Cathode Ray Tube
CT	Magnet, A Current Transformer (type of magnet)
CT	Hall Effect, A Current Transformer Based on the Hall Effect
CVT	Continuously Variable Transmissions
D/A	Digital-to-Analog Converter
DC	Direct Current
DI	Direct Injection
DMM	Digital Multimeter
DSO	Digital Storage Oscilloscope
DSP	Digital Signal Processor
DTC	Dead Top Center, and this term also means Diagnostic Error Code
DVOM	Digital Volts-Ohm Meter, the same as a DMM
ECT	Engine Coolant Temperature
EGO	Exhaust Gas Oxygen
EGR	Exhaust Gas Recirculation

EI	Electronic Ignition
EMI	Electromagnetic Interference, also called RFI or Radio Frequency Interference
EOC	End of Conversion
EST	Electronic Spark Timing
EVP	EGR Valve Position
FFCS	Fuel Flow Control System, also called the Feedback Control System
FT	Fuel Temperature
HC	Hydrocarbon Chemicals
IAT	Intake Air Temperature
IC	Integrated Circuit
KEKO	Key On - Engine Off condition
LCD	Liquid Crystal Display
LED	Light Emitting Diode
LSI	Large Scale Integration
MAF	Mass Airflow (a very important type of sensor)
MAP	Manifold Absolute Pressure (Another very important type of sensor)
MBT	Mean Best Torque
MFI	Multiport Fuel Injection
MIL	Malfunction Indicator Lamp
MPU	Microcomputer Processing Unit
MSI	Medium Scale Integration
MUX	Multiplexer
NiCad	(Nickel Cadmium) A type of rechargeable battery
NOX	Various Combinations of Oxides in Nitrogen
O_2	Oxygen (diatomic) the kind we breathe
O_2FB	Oxygen Feedback Sensor
OBDII	On-Board Computer Diagnostics II
PCB	Printed Circuit Board
PCM	Power Train Control/Management Module
PI	Proportional Integral

PID	Proportional Integral Derivative	SSI	Small Scale Integration
PIP	Profile Ignition Pickup	TBI	Throttle Body Injector
POT	An abbreviation for potentiometer, which is a variable resistor	TDC	Top-Dead-Center
		TWC	Three-Way Catalyst
PSI	Pounds Per Square Inch	VIAS	Vehicle Impedance Aspiration System
PSIG	Pounds Per Square Inch Gage		
RAM	Random Access Memory	VF	Vacuum Fluorescent Display, or also Voltage-to-Frequency converter
ROM	Read Only Memory		
RPM	Revolutions Per Minute		
SDSG	Silicon Diffused Strain Gage	VLSI	Very Large Scale Integration
SP	Stack Pointer	VSS	Vehicle Speed Sensor
SPOUT	Spark Output	WOT	Wide Open Throttle

Application Guide

	Amps DC*	Analog Pointer	ALERT	Continuity	% DUTY	Hz	Lo-Ohms**	Milliamps	Millivolts	MIN MAX	Ohms	ms-PULSE	RPM	HOLD	Volts AC	Volts DC	ZERO Δ
IGNITION/ENGINE																	
Coils						•				•						•	
Computer Temp Sensors									•	•						•	•
Condensers (Capacitors)		•								•						•	
Connectors		•	•			•		•	•	•			•			•	
Contacts Set		•		•	•	•		•		•						•	
Distributor Cap										•							
Engine Speed													•				
Feedback Carburetors					•	•				•				•	•	•	
Fuel Injectors (Electronic)	•				•	•				•	•					•	
Hall-Type Sensors	•					•			•	•		•		•	•	•	
Idle Air Motors	•				•	•	•	•		•	•						
Ignition Modules	•					•										•	
MAF Sensor						•			•				•				
Magnetic Pickups	•		•		•	•		•	•	•			•		•	•	
MAP & BP Sensors	•					•				•						•	
O₂ Sensors	•					•			•	•							
Throttle Position Sensors	•					•			•	•						•	•
STARTING SYSTEM																	
Battery	•					•			•							•	
Connectors		•				•		•		•			•		•	•	•
Interlocks (neutral safety switch)		•	•			•			•	•						•	
Solenoids			•			•	•	•	•	•						•	
Starters	•							•	•				•		•	•	•
COOLING SYSTEM																	
Connectors		•	•					•	•	•			•			•	
Fan Motor		•								•						•	
Radiator									•								
Relays		•	•			•			•	•						•	
Temperature Sensors		•							•	•							
Temperature Switches										•		•	•	•	•	•	
CHARGING SYSTEM																	
Alternators	•			•		•			•	•					•	•	
Computerized Regulators	•				•	•				•			•				
Connectors		•	•			•		•	•	•			•				
Diodes, (AC Ripple)				•					•	•				•			
Diode Rectifier		•	•													•	
Regulators	•	•				•			•	•						•	
BODY ELECTRIC																	
Compressor Clutch		•	•			•			•	•						•	
Lighting Circuits		•	•			•				•						•	
Relay and Motor Diodes		•		•													
Transmissions		•				•			•	•							

* Used with Fluke 80i-410 or 80i-1010 current clamp.
** Available in Fluke 88 only.

Table of Equivalents

These equivalents are useful in establishing "unit ratios." All underlined figures are exact.

Units of Length	1 inch = <u>2.54</u> centimeter = <u>25.4</u> mm 1 mile = 1.609 kilometer 1 microinch (μin.) = 10^{-6} in.	1 foot = <u>0.3048</u> meter 1 mile = <u>5280</u> ft 1 μm = 10^{-6} m = 10^{-4} cm
Units of Area	1 square inch = 6.45 cm^2 1 square mile = 2.59 km^2	1 square foot = 929 cm^2 1 acre = <u>43 560</u> ft^2
Units of Volume	1 gallon (U.S.) = <u>231</u> in.^{3} 1 gallon = 3.79 liter 1 gallon water = 8.34 lb @ 60° 1 in.^{3} = 2.54^3 cm^3 = 16.4 cm^3	1 gallon = 4 quarts = 8 pints 1 liter = 1000 milliliter 1 L = 1000 cm^3 (cc)
Force and Mass	1 newton = 0.102 kg force (wt) = 0.225 lb 10^{-5} newton = 1 dyne 1000 kg = 1 metric ton = 1.103 English tons	1 pound = 16 oz 1 ton = 2000 lb 1 slug force (wt) = 32.2 lb
Speed	60 mi/hr = 88 ft/s	1 mi/hr = 1.609 km/hr
Angle	1 revolution = 360° = 2π rad	1 radian (rad) = 57.3°
Temperature	°Fahrenheit = 1.8°C + 32 °Rankine = °F + 460	°Celsius = (°F − 32)/1.8 Kelvin = °C + 273
Pressure	1 newton/m^2 = 1 pascal (Pa) 1 atmosphere = 76 cm Hg = 29.95 in. Hg = 14.7 lb/in.^{2} This conversion can be used to form unit ratios, i.e., $$\frac{76 \text{ cm Hg}}{29.95 \text{ in. Hg}} = 1$$	6890 pascal = 1 lb/in.^{2}
Computer Conversions	8192 bits = 1 K (kilobyte) RAM $\dfrac{1 \text{ page}}{\text{minute}} = \dfrac{25 \text{ characters}}{\text{second}}$ (printer speed based on 60 characters/line and 25 lines/page)	1 screen pixel = at least 1 bit of memory

Engineering Prefixes

Number	Power of Ten	Prefix	Symbol
1 000 000 000	10^9	giga-	G
1 000 000	10^6	mega-	M
1 000	10^3	kilo-	k
1	10^0	unity	none
0.001	10^{-3}	milli-	m
0.000 001	10^{-6}	micro-	μ
0.000 000 001	10^{-9}	nano-	n
0.000 000 000 001	10^{-12}	pico-	p

Index

Index

D

E

T

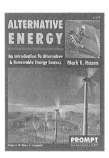

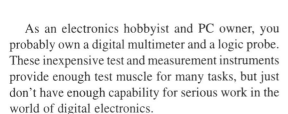

PC Hardware Projects, Volume 3
James "J.J." Barbarello

PC Hardware Projects, Volume 3 shows you how to construct ComponentLAB, a device that connects to the digital I/O card of previous volumes and gives you the capability to measure and record DC voltage, resistance and capacitance, and test digital ICs. ComponentLAB also contains an 8-bit A/D converter that you can use to capture any analog signal ranging from a few hundred millivolts all the way up to 5 volts. This book also explains how to export the data you store into other applications for your own data processing needs, and provides you with all the inside information on how to use the digital I/O card for other purposes.

PC Hardware Projects, Volume 3 comes with a disk containing the application programs for ComponentLAB, the A/D converter, and the digital I/O card. It also contains definition files for 95 of the most common TTL and CMOS ICs, and various sample data files.

Projects
204 pages • paperback • 7-3/8 x 9-1/4"
ISBN: 0-7906-1151-1 • Sams: 61151
$29.95

Real-World Interfacing With Your PC, *Second Edition*
James "J.J." Barbarello

As the computer becomes increasingly prevalent in society, its functions and applications continue to expand. Modern software allows users to do everything from balance a checkbook to create a family tree. Interfacing, however, is truly the wave of the future for those who want to use their computer for things other than manipulating text, data, and graphics.

Real-World Interfacing With Your PC provides all the information necessary to use a PC's parallel port as a gateway to electronic interfacing. In addition to hardware fundamentals, this book provides a basic understanding of how to write software to control hardware. While the book is geared toward electronics hobbyists, it includes a chapter on project design and construction techniques, a checklist for easy reference, and a recommended inventory of starter electronic parts to which readers at every level can relate.

Projects
120 pages • paperback • 7-3/8 x 9-1/4"
ISBN: 0-7906-1145-7 • Sams: 61145
$29.95

**To order your copy today or locate your nearest Prompt®
Publications distributor : 1-800-428-7267 or www.hwsams.com**

Prices subject to change.

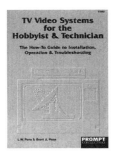

Advanced Electronic Projects For Your Home and Automobile, *2nd Ed.*

Stephen Kamichik

You will gain valuable experience in the field of advanced electronics by learning how to build interesting and useful projects in this book. You can build the projects covered in *Advanced Electronic Projects, Second Edition* whether you are an experienced electronics hobbyist or an electronics engineer. Everything in this book has been thoroughly tested by the author, and a detailed explanation of each circuit is given to help you understand its operation. Each project is designed to give you years of enjoyment and reliable service.

Learn how to construct: Active Filters, Stereo Preamplifier, Speaker Systems, Bipolar Power Supplies, Infrared Remote Control, DC Motor Speed Control, Electronic Scorekeeper, Brain Wave Monitor, Car Anti-Theft System, and More!

Projects
144 pages • paperback • 6 x 9"
ISBN: 0-7906-1161-9 • Sams 61161
$24.95

Fun Projects for the Experimenter

Newton C. Braga

Author Newton C. Braga, whose works have appeared in electronics magazines for over 20 years, has collected fifty of his most fun, easy-to-build, and practical projects for your enjoyment. Basic electronic principles and fundamentals are stressed. These projects are primarily stand-alone, low-cost, and with few components, intended for one evening of work. The components needed are listed along with schematics, and hints and questions about the circuits are included to stimulate your imagination regarding possible modifications and alternate use.

Examples of the projects include an LED flasher, mini-metronome, electronic fishing lure, micro FM transmitter, touch switch, wireless beeper, and signal tracer. For hobbyists and students wanting to understand the basic principles of electronics, this book will provide answers to many of your questions.

Projects
328 pages • paperback • 7-3/8 x 9-1/4"
ISBN: 0-7906-1149-X • Sams 61149
$24.95

To order your copy today or locate your nearest Prompt® Publications distributor : 1-800-428-7267 or www.hwsams.com

Prices subject to change.